# INTEGRATION
# IN FINITE TERMS

# INTEGRATION IN FINITE TERMS
## Liouville's Theory of Elementary Methods

### Joseph Fels Ritt
Davies Professor of Mathematics
Columbia University

New York
Columbia University Press
1 9 4 8

COPYRIGHT 1948 COLUMBIA UNIVERSITY PRESS, NEW YORK

Published in Great Britain and India
by Geoffrey Cumberlege
Oxford University Press, London and Bombay

MANUFACTURED IN THE UNITED STATES OF AMERICA

# PREFACE

During the period between 1833 and 1841, J. Liouville presented a theory of integration in finite terms. He determined the form which the integral of an algebraic function must have when the integral can be expressed with the operations of elementary mathematical analysis, carried out a finite number of times. He showed that the elliptic integrals of the first and second kinds have no elementary expressions. He proved that certain simple differential equations cannot be solved by elementary procedures. His papers contain other remarkable applications of his theory.

The questions treated by Liouville are questions which occur to every strong undergraduate student of mathematics. Nevertheless Liouville's work never received very wide attention. It has always been something which everyone would like very much to know about but which very few undertake to study.

During the nineteenth century, extremely little was done in direct continuation of Liouville's work.* About forty years ago, the Russian mathematician Mordukhai-Boltovskoi began to write on Liouville's theory and contributed extensively to it. In particular, he published a book on the integration of transcendental functions and one on the integration in finite terms of linear differential equations. Through his influence, the subject seems to have been more widely studied in Russia than elsewhere. The present writer published some work on these questions between 1923 and 1927. At the present time, Ostrowski is writing on the subject.

This monograph gives an account of Liouville's work and of some of that of his few followers. On the basis of what has already been said, a glance through the chapters, or even over the table of contents, will give a sufficient idea of the topics covered.

I should like, however, to say something in regard to the treatment given here of Liouville's work. Liouville's methods are ingenious and beautiful. From the formal standpoint, they are entirely sound. There are, however, certain questions connected with the

---

* To be sure, there appeared the Picard-Vessiot theory of linear differential equations, which furnishes, for such equations, results analogous to those of Galois for algebraic equations. Recent work of E. R. Kolchin has brought rigor and simplicity to the Picard-Vessiot theory.

One should perhaps mention also the remarkable work of Bruns on the algebraic solutions of the equations of celestial mechanics. (*Acta Mathematica*, Vol. XI.)

many-valued character of the elementary functions which could be pressed back behind the symbols in Liouville's time but which have since learned to assert their rights. Such matters are mulled over in the first chapter. The mulling is inescapable. It might be great fun to talk just as if the elementary functions were one-valued. I might even sound convincing to some readers; I certainly could not fool the functions. However, if one is chiefly interested in formal ideas, one may read the "First Survey of the Elementary Functions" in Chapter I and then pass to the summary at the end of that chapter. It ought not to be hard, after that, to follow the formal processes of Chapter II.

As regards the theory of functions, I have assumed, in Chapter I, an acquaintance with the simpler facts concerning analytic continuation. Riemann surfaces of algebraic functions are mentioned in Chapter I; their simpler properties permit, in Chapter II, the swift liquidation of questions on the integrability of special algebraic functions. In algebra, I use in Chapter I the discriminant of a polynomial and the resultant of a pair of polynomials. For the rest, special material of algebra or of analysis is developed wherever it is needed.

I should like, in conclusion, to say something concerning Joseph Liouville (1809-1882). He originated the notion of derivative of fractional order. The first examples of transcendental numbers were due to him; his work on this question was the starting point of the modern researches of Thue and Siegel. He was one of the founders of the theory of boundary value problems. As Abel had also done, earlier, he solved a problem involving an integral equation; this was decades before the general theory of such equations came into being. His work on doubly periodic meromorphic functions was precursive to Weierstrass's theory of the elliptic functions, just as his work in the theory of elimination anticipated, to some extent, ideas of Kronecker. He presented a method for treating classes of Diophantine equations which has been developed extensively, in recent times, by E. T. Bell. In geometry, he determined the group of conformal transformations in three dimensions. He was the founder of the *Journal de mathématiques pures et appliquées*. Considering his achievements, one may question whether he has been adequately appreciated, even by the mathematicians of his own country. It is surprising, for instance, that his collected works were never published.

## PREFACE

If this monograph should promote the study of what is probably one of the most interesting portions of Liouville's work, the writer will feel amply rewarded.

J. F. Ritt

*Columbia University*
*July, 1947*

## CONTENTS

| | | |
|---|---|---|
| I. | Elementary Functions of One Variable | 3 |
| II. | Algebraic Functions with Elementary Integrals | 20 |
| III. | Integration of Transcendental Functions | 40 |
| IV. | Further Questions on the Elementary Functions | 53 |
| V. | Series of Fractional Powers | 61 |
| VI. | Integration of Differential Equations by Quadratures | 69 |
| VII. | Implicit and Explicit Elementary Solutions of Differential Equations of the First Order | 78 |
| VIII. | Further Implicit Problems | 87 |
| | References | 99 |

# INTEGRATION
# IN FINITE TERMS

## Chapter I
## ELEMENTARY FUNCTIONS OF ONE VARIABLE

*FIRST SURVEY OF THE ELEMENTARY FUNCTIONS*

1. The functions which we shall study in the present chapter are essentially those which make up the functional world of a student of the integral calculus. Such a student, if not familiar with the concept of algebraic function in its most general form, knows the polynomials and fractional rational functions, has seen functions involving radicals, and can imagine quite well the most general algebraic function which can be expressed in terms of radicals. He knows $e^x$, log x, sin x, cos x, and the inverses of the latter two functions. After compounding functions of the foregoing types in various ways to produce combinations like $e^{x^2}$ or log (sec x + tan x), he is in possession of an extensive class of functions, each constructed with a finite number of operations. A typical example would be

$$\tan [e^{x^2} - \log_x (1 + x^{\frac{1}{2}})] + [x^x + \log \arcsin x]^{\frac{1}{2}}.$$

The expression $x^x$ is to be interpreted, of course, as $e^{x \log x}$. As to the logarithm to the base x, it is nothing more than the natural logarithm divided by log x.

The use of only a finite number of operations needs particular emphasis. One meets infinite series in a calculus course. The representation of functions by means of such series is a question foreign to our present study. We shall be concerned only with what can be obtained from the basic functions with calculations involving only a finite number of operations.

2. An inspection of our functions will permit us to describe them more closely and to classify them. We shall look twice at our material, first casually, just to see how matters stand, then fully and squarely, with no turning away from hard realities.

We notice first that, if complex numbers are employed, the trigonometric functions and their inverses become redundant. For instance

$$\sin x = \frac{e^{ix} - e^{-ix}}{2i}$$

and

$$\arcsin x = \frac{1}{i} \log (ix + \sqrt{1 - x^2}).$$

Thus the functions of elementary analysis are constructed out of the variable x by repeated use of the following operations:
(a) Algebraic operations performed on one or more expressions
(b) The taking of exponentials
(c) The taking of logarithms.
Just what is to be understood by an algebraic operation will be explained fully later.

Now let us explain how the elementary functions will be classified. The variable x will be called a *monomial of order zero*, and any algebraic function of x will be called a *function of order zero*. The exponential or the logarithm of a nonconstant algebraic function will be called a *monomial of the first order*. An algebraic combination of x and of monomials of the first order will, if it is not an algebraic function, be called a *function of the first order*. The exponential or the logarithm of a function of the first order, will, if it is not a function of order zero or a function of order unity, be called a *monomial of the second order*. For instance, as one will see in Chapter IV, the exponential of $e^x$, and log (log x) are monomials of the second order. An algebraic combination of monomials of orders 0, 1, 2 which is not a function of one of the orders 0, 1, will be called a *function of the second order*. The classification continues in this way.

All of this needs a closer examination, and to that we now turn.

## *ALGEBRAIC FUNCTIONS*

3. Let us recall the notion of algebraic function. A function u of x is algebraic if it is defined by an irreducible relation

(1) $$\alpha_0(x) \, u^p + \alpha_1(x) \, u^{p-1} + \ldots + \alpha_p(x) = 0$$

with $p > 0$ where the $\alpha$ are polynomials in x with any complex numbers for coefficients, $\alpha_0$ not being identically zero. In saying that (1) is irreducible, we shall mean that its first member is not the product of two nonconstant polynomials in u and x.

We shall find ourselves at times employing the Riemann surface of u, which is a surface of p sheets. For our immediate purposes, it is more important to consider u as a *monogenic analytic function* (m.a.f.) in the sense of Weierstrass, that is, as the totality of power series which can be obtained from some given power series by analytic continuation. Each power series is called an element of the m.a.f. An element which is a series of powers of $x - x_0$ may be represented by $P(x-x_0)$.

Let $P(x-x_0)$ be any element of our algebraic function u. Let C be any curve with $x_0$ for first point. We may consider C to be given by a relation

(2) $$x = \varphi(\lambda)$$

where $\varphi(\lambda)$ is a continuous, complex-valued function of the real variable $\lambda$ on the closed interval (0, 1), with $\varphi(0) = x_0$.

If $P(x-x_0)$ cannot be continued along the entire curve C, it must be that C has on it one or more points which are places at which u has poles or branch points. If that is the case, we can, by bending C slightly so as to avoid the singularities of u, which are finite in number, obtain a new curve, along all of which u can be continued. This simple step of bending a curve so as to permit the continuation of an element leads to a general idea which will help us considerably in the study of the elementary functions.

## FLUENT FUNCTIONS

4. An analytic function of x will be said to be *fluent* if every element $P(x-x_0)$ of the function has the following property. *For every curve (2) with* $\varphi(0) = x_0$, *and for every* $\varepsilon > 0$, *there can be found a curve*

(3) $$x = \varphi_1(\lambda) \qquad (0 \leq \lambda \leq 1)$$

*with* $\varphi_1(0) = x_0$, *such that*

$$|\varphi_1(\lambda) - \varphi(\lambda)| < \varepsilon$$

*for* $0 \leq \lambda \leq 1$ *and such that* $P(x-x_0)$ *can be continued along the entire curve (3)*. Broadly speaking, an element of the function, if it cannot be continued along a given path, can be continued along some path in every neighborhood of the given one.

As we have indicated, algebraic functions are fluent. To be formal, one replaces (2) by a curve consisting of short straight segments and then uses arcs of small circles to avoid the finite number of singular points which may lie on the segments.

## EXPONENTIALS OF ALGEBRAIC FUNCTIONS

5. Let v be a nonconstant algebraic function. We are interested in the nature of $e^v$. We consider any element of v and take its exponential. This furnishes an analytic element, the totality of whose continuations constitutes $e^v$. This process furnishes a unique m.a.f., for if $P_1$ and $P_2$ are elements of v, the path which continues $P_1$ into $P_2$ continues the exponential of $P_1$ into that of $P_2$. The

Riemann surface of $e^v$ is that of v. Our definition of $e^v$ gives $e^v$ a value at each finite point of the surface which is not a branch point and at which v has no pole. It is possible to go further. At a branch point of v which is not a pole, corresponding to a value $x_0$ of x, we have merely to define $e^v$ as the exponential of the value of v. We secure an expansion of $e^v$ for the branch point, in fractional powers of $x - x_0$, by taking the exponential of the expansion of v. If v has a pole at a point of the surface with $x = x_0$, we secure for $e^v$ an expansion containing an infinite number of negative powers of $x - x_0$, fractional powers appearing in the expansion if we are dealing with a branch point. If $x_0$, in what precedes, is $\infty$, we obtain for $e^v$ an expansion in integral or fractional powers of $1/x$.

Two distinct elements $P_1(x-x_0)$ and $P_2(x-x_0)$, formed for the same point $x_0$, produce two distinct elements $Q_1$ and $Q_2$ of $e^v$. If $Q_1$ were identical with $Q_2$, we would have

$$P_2 = P_1 + 2k\pi i$$

with k an integer distinct from 0. The path which continues $P_1$ into $P_2$ would thus continue $P_2$ into $P_3 = P_2 + 2k\pi i$. The same path would continue $P_3$ into a new element. Continuing, we would find an infinite number of elements for v at $x_0$, a contradiction of the algebraic nature of v.

In passing, let us notice that if v is not a constant, $e^v$ is not algebraic. If it were, $e^x$, resulting from $e^v$ by an algebraic substitution for x, would be algebraic.* The nonalgebraic character of $e^x$ follows from the fact that it assumes the same value for an infinite number of values of x, something which cannot happen for an algebraic function, as can be seen from (1), in which a value assigned to u leads to an algebraic equation for x.

That $e^v$ is fluent follows immediately from the fact that v is.

## LOGARITHMS OF ALGEBRAIC FUNCTIONS

6. We now examine log v, where v is algebraic and not constant. It will turn out that there may be more than one m.a.f., but always a finite number of such functions, which it is proper to call log v.

Let $P(x-x_0)$ be an element of v, for some $x_0$, which is not zero at $x_0$; that is, the series P contains a constant term distinct from 0. We can form an infinite number of elements log P, any

---
* Such questions of substitution will receive formal treatment later.

two differing by an integral multiple of $2\pi i$. Let Q be any of these elements. We obtain from Q, by continuation, an m.a.f. which we shall call log v.

An example will show that distinct elements Q may lead to distinct functions log v. Let $v = x^2$ and let $x_0$ be any complex number distinct from 0. Let a Q be selected. Let x describe a closed path, beginning and ending at $x_0$, which does not pass through the origin. As x travels over this path, its amplitude changes by a multiple of $2\pi$, so that the amplitude of $x^2$ changes by an even multiple of $2\pi$. As Q is continued along the path, its coefficient of i, which is one of the amplitudes of $x^2$, changes by an even multiple of $2\pi$. Thus there is no path which continues Q into $Q + 2\pi i$. It is easy to see that there are just two functions log $x^2$, the doubles of log x and log (-x).

In the general case, there is no difficulty in showing that there is more than one log v only when v is of the form $u^r$ with r an integer greater than unity and u a function which is one-valued on the surface of v. If r is taken as large as possible, there are just r functions log v.

All elements $P(x-x_0)$ lead to the same functions log v, since a Q secured from any element $P_1$ can be continued, along a suitable path, into a Q for any other element $P_2$.

Thus, given v, we must, in referring to log v, specify one of a set of functions, any two of which differ by a constant. It is easy to see that each log v is fluent; given a path (2) we can bend it slightly to avoid the singularities and zeros of v.

## ALGEBRAIC COMBINATIONS

7. Let $\mathcal{E}_1, \ldots, \mathcal{E}_r$ be monogenic analytic functions, not necessarily distinct from one another. Let $x_0$ be any point at which each $\mathcal{E}_i$ has at least one analytic branch and let $P_i(x-x_0)$, $i = 1, \ldots, r$, be an element of $\mathcal{E}_i$. Let $u(x)$ be an m.a.f. for which some $x_0$ exists as above at which u has an element which satisfies an equation

(4)  $$\alpha_0 u^p + \ldots + \alpha_p = 0$$

where each $\alpha$ is a polynomial, with constant coefficients, in the $P_i(x-x_0)$, and where $\alpha_0$ is not identically zero. We shall call $u(x)$ an *algebraic combination* of the $\mathcal{E}_i$.

Thus the integral function

which may be written $e^{\sqrt{x}} + e^{-\sqrt{x}}$ may be regarded as an algebraic combination of the two identical monomials $e^{\sqrt{x}}, e^{\sqrt{x}}$. We may use any $x_0 \neq 0$ and take $P_1$ and $P_2$ as the two elements of $e^{\sqrt{x}}$ at $x_0$. One need not fear that the definition of algebraic combination makes the symbol $e^{\sqrt{x}} + e^{-\sqrt{x}}$ ambiguous. Our definition of algebraic combination does not make it necessary to break with established notation and to create a symbolism which will describe everything involved in making algebraic combinations.

## FUNCTIONS OF FIRST ORDER

8. If v is a nonconstant algebraic function, each of the monogenic analytic functions $e^v$ and log v will be called a *complete monomial of order 1*. The variable x will be called a complete monomial of order 0. By a *monomial* we shall mean a branch of a complete monomial which is analytic in some area.

An m.a.f. $u(x)$ will be called a *function of order 1* if it is not algebraic and if it is an algebraic combination of complete monomials of orders 0 and 1. Frequently we shall use the term *function of order 1* to represent a branch of a function of order 1.

Let u be a function of order 1. We are going to prove that u is fluent. We start by considering the *field* $\mathcal{D}$ of x and the $P_i$ which figure in the equation (4) defining u. $\mathcal{D}$ is the totality of rational combinations, with complex coefficients, of x and the $P_i$. The functions in $\mathcal{D}$ are all meromorphic in some circle with center at $x_0$. We may assume the first member of (4) to be irreducible in $\mathcal{D}$. If it is not, we replace it by that one of its irreducible factors which vanishes for the given element of u. Then the discriminant D of (4), which is a polynomial in the coefficients in (4) and is a function analytic at $x_0$, cannot vanish identically. If it did, the first member of (4) would have a factor of positive degree in common with its derivative with respect to u; this is because the resultant of the first member of (4) and its derivative equals $\alpha_0 D$ or its negative.

Now let us see what will assure the possibility of continuing u along a curve C which starts at $x_0$. Suppose that each $P_i$ in (4) can be continued along C. This means that $\alpha_0$ and the discriminant can be continued along C. Suppose now that neither $\alpha_0$ nor the discriminant vanishes at any point of C. Then (4), considered an algebraic equation for a quantity u, will provide p distinct values for u at each point of C. These values will furnish p distinct functions, each analytic along C. One of these functions will effect the continuation along C of the given element of u at $x_0$. If C does not satisfy the foregoing conditions, we first bend it slightly so as to have a curve C' along which each P can be continued. This is possible because each complete monomial is fluent. It may be that $\alpha_0$ or the discriminant has zeros along C'. Let us consider, for instance, $\alpha_0$. The continuation of $\alpha_0$ along C' is effected by the use of a finite set of elements. Each of these elements represents an analytic function, the zeros of which are naturally isolated. Thus a slight bending of C' will give us a curve along which $\alpha_0$ is not zero, except perhaps at $x_0$, which is the only point being held fast. Now if $\alpha_0$ is zero at $x_0$, we can start operations from a point on C close to $x_0$ at which $\alpha_0$ is not zero. Similar remarks apply to the discriminant.

Thus a function of the first order is fluent.

9. Now let $\mathfrak{U}$ be any area in the complex plane and suppose that we can continue the above-mentioned element of u with center at $x_0$ into and all over $\mathfrak{U}$, so that u has a branch which is uniform and analytic throughout $\mathfrak{U}$. Let C be some curve along which u can be continued from $x_0$ into $\mathfrak{U}$. Bending C slightly if necessary, we may suppose that each P can be continued along C. Thus, in every area $\mathfrak{U}$ in which u is analytic, there is an area $\mathfrak{U}_1$ in which the complete monomials $\mathcal{E}$ of which the P are elements have analytic branches $\theta_i(x)$, $i = 1, \ldots, r$ and in which there is satisfied an equation

$$(5) \qquad \beta_0 u^p + \ldots + \beta_p = 0$$

with each $\beta_i$ the same polynomial in x and the $\theta(x)$ which $\alpha_i$ is in x and the P. We may take $\mathfrak{U}_1$ so that each $\theta_i(x)$ is an exponential or a logarithm of a branch, analytic in $\mathfrak{U}_1$, of the algebraic function of which $\mathcal{E}_i$ is an exponential or a logarithm.

Suppose that in (4) we replace each $P_i$ by a variable $x_i'$. Replacing u by a variable v, we form an equation

$$(6) \qquad \alpha_0(x, x_i')v^p + \ldots + \alpha_p(x, x_i') = 0.$$

Let C be a curve, joining $x_0$ to a point $\underline{a}$, along which u and the P can be continued. C continues each $P_i$ into a function $\theta_i(x)$ analytic in a neighborhood of $\underline{a}$. Suppose that $\alpha_0$ in (4) and the discriminant of (4), both continued along C, are not zero at a. Then, for $x = a$, $x_i' = \theta_i(a)$, $i = 1, \ldots, r$, the first coefficient in (6) and the discriminant of (6) do not vanish. This means that for the same values of x and the x', and for v equal to u(a), the derivative with respect to v of the first member of (6) does not vanish. Using the implicit function theorem on (6), we can solve for v, obtaining an algebraic function of x and the x' which is analytic when those variables remain in the neighborhood of $x = a$, $x'_i = \theta_i(a)$ and which, when each $x'_i$ is replaced by $\theta_i(x)$, reduces, in a neighborhood of a, to the continuation of $u(x)$ along C.

10. On the basis of §9, the structure of a function u of order 1 may be described as follows. Given any area in which some branch of u is analytic, there exist

(0) a point $\underline{a}$ interior to the area, a $\rho > 0$ and a $\rho_1 > \rho$;

(I) r algebraic functions of x, each analytic for $|x - a| < \rho_1$;

(I') r monomials $\theta_1, \ldots, \theta_r$, each an exponential or a logarithm of one of the functions in (I), each analytic for $|x - a| < \rho$ and such that $|\theta_i(x) - \theta_i(a)| < \rho_1$ for $|x - a| < \rho$, $(i = 1, \ldots, r)$;

(II) an algebraic function of the variables $x, x'_1, \ldots, x'_r$ which is analytic for $|x - a| < \rho_1$, $|x'_i - \theta_i(a)| < \rho_1$, $(i = 1, \ldots, r)$ and which reduces to the given branch of u for $|x - a| < \rho$ if each $x'_i$ is replaced by $\theta_i(x)$.*

Furthermore, the integer r, the algebraic equations satisfied by the functions in (I) and that in (II), and the exponential or logarithmic characters of the $\theta_i$ are independent of the area in which u is considered and of the branch of u.

The first member of (6) is irreducible as a polynomial in v, x, and the x'. This does not mean that when the x' are replaced by their monomials we obtain an equation (4) which determines a single analytic function u. For instance,

$$u^2 - e^x = 0$$

gives two distinct analytic functions. There is no point of difficulty here. A function u of the first order is a definite function, for which a scheme of construction can be given as above. There may be other functions whose schemes of construction employ the same material which appear in the scheme for u.

## FUNCTIONS OF ANY ORDER

11. We can now define, by induction, functions of any order n. The exponential or a logarithm of a function of order n-1 will be called a *complete monomial of order n*, provided that it is not among the functions of orders 0, 1, ..., n-1. A branch of a complete monomial, analytic in some area, will be called a monomial.

---

* One may ask why, in arranging our material, we do not first choose $\rho_1$ and then a sufficiently small $\rho$, without requiring that $\rho < \rho_1$. The reason is that the algebraic function in (II) may depend effectively on x.

ELEMENTARY FUNCTIONS    11

By a *function of order n*, we shall mean an algebraic combination of monomials of orders $0, 1, \ldots, n$ which is not a function of one of the orders $0, 1, \ldots, n-1$. A branch of such an m.a.f. will also called, at times, a function of order n.

As above, we may assume that the discriminant of the equation like (4) which describes the algebraic combination does not vanish identically.

The existence of functions of all orders will be seen in Chapter IV. One sees by a quick induction that a function of any order n is fluent.

The structural scheme of a function of order n is built up by induction, with the help of the notion of fluency. Given a function u of order n, analytic in some area, there exist

(0) a point $\underline{a}$ interior to the area, a $\rho > 0$ and a $\rho_1 > \rho$;

(I) $r_1$ algebraic functions of x, each analytic for $|x - a| < \rho_1$;

(I') $r_1$ monomials $\theta_1', \ldots, \theta_{r_1}'$, each either an exponential or a logarithm of one of the functions in (I), each analytic for $|x - a| < \rho$ and such that $|\theta_i'(x) - \theta_i'(a)| < \rho_1$, for $|x - a| < \rho$, $(i = 1, \ldots, r_1)$;

(II) $r_2$ algebraic functions of x and of $r_1$ other variables $x_1', \ldots, x_{r_1}'$, each analytic for $|x - a| < \rho_1$, $|x_i' - \theta_i'(a)| < \rho_1$;

(II') $r_2$ monomials $\theta_1'', \ldots, \theta_{r_2}''$, each an exponential or a logarithm of one of the functions of order 1 to which the algebraic functions in (II) reduce when each $x_i'$ is replaced by $\theta_i'$; each $\theta_i''$ is analytic for $|x - a| < \rho$ and $|\theta_i''(x) - \theta_i''(a)| < \rho_1$ for $|x - a| < \rho$;

. . . . . . . . . . .

(N + 1) an algebraic function of x; $\ldots$; $x_1^{(n)}, \ldots, x_{r_n}^{(n)}$, analytic for $|x - a| < \rho_1, \ldots, |x_1^{(n)} - \theta_1^{(n)}(a)| < \rho_1$, which reduces to the given branch of u when each variable is replaced by the monomial which corresponds to it.*

Furthermore, the integers $r_i$, the algebraic equations satisfied by the functions in (I), (II), $\ldots$, (N + 1), and the character of the $\theta$ as exponentials or logarithms are independent of the area in which u is considered and of the branch of u.

A function u of order n, described as above at a point $\underline{a}$, will be said to be of *regular structure* at a.

We see that an accented x may be used in forming a monomial of higher order than that to which it corresponds, and may be used

---

* Of course, x itself is not actually replaced.

again by itself. Such would be the case, for instance, in the scheme of structure of

$$(\log \log x) \log x.$$

We have chosen a symbolism which allows this, for the purposes of our work on integration. To hint at what is involved here, let us consider the function

$$u = (\log \log x)^2$$

in which log x is used only to build a higher monomial. In the derivative of u, the monomial log x appears by itself, as well as in log log x.

A monomial of order n will frequently be described as an *n-monomial* and indeed as an *n-exponential* or an *n-logarithm*, according to its character.

The functions to which orders are assigned by the preceding definitions will be called *elementary functions of x*.

As Bieberbach has pointed out,* the fluency of the elementary functions makes visible immediately the nonelementary character of a function which has a curve made up of singularities, for instance, a natural boundary. Thus many of the higher automorphic functions, functions which satisfy simple algebraic differential equations, are not elementary. The integrals of algebraic functions and the elliptic functions cannot be treated in this way.

### ALGEBRAIC COMBINATIONS OF ELEMENTARY FUNCTIONS

12. We shall now inquire into the nature of algebraic combinations of elementary functions. To get an idea of what is in this question, let us consider the process of forming a function u of order 2. One would use an equation (4) in which the $\alpha$ are polynomials in elements of complete monomials of orders 0, 1, 2. It is natural to ask what would happen if, with greater apparent generality, one permitted the $\alpha$ to be polynomials in complete monomials of order 2 and in elements of *any functions* of orders 0, 1. The answer is that nothing new would be obtained; the use of monomials is sufficient. This will be seen after we have obtained a set of theorems that parallel those theorems in the theory of algebraic numbers which show that the system of algebraic numbers is closed with respect to algebraic operations.

In what follows, all functions will be analytic in some area and n will be a given positive integer.

* *International Congress of Mathematicians* (Zurich, 1932), 1, 164.

LEMMA: *Let $\varphi$ be an analytic function. Let there exist, for some $k \geq 1$, k analytic functions $\xi_1, \ldots, \xi_k$, not all zero, such that*

(7)
$$\varphi \xi_1 = a_{11} \xi_1 + \ldots + a_{1k} \xi_k$$
$$\cdot \quad \cdot \quad \cdot \quad \cdot \quad \cdot \quad \cdot$$
$$\varphi \xi_k = a_{k1} \xi_1 + \ldots + a_{kk} \xi_k$$

*where the $a_{ij}$ are polynomials in monomials of the first n orders. Then $\varphi$ is elementary and its order does not exceed n.*

The equations (7) are a system of homogeneous linear equations in the $\xi$ which have a solution with some $\xi$ not zero. The determinant

$$\begin{vmatrix} \varphi - a_{11}, & -a_{12}, & \ldots, & -a_{1k} \\ \cdot & \cdot & \cdot & \cdot \\ -a_{k1}, & -a_{k2}, & \ldots, & \varphi - a_{kk} \end{vmatrix}$$

must vanish. This gives an equation

$$\varphi^k + c_1 \varphi^{k-1} + \ldots + c_k = 0$$

where the c are polynomials in monomials of orders $0, \ldots, n$. Thus, $\varphi$ is elementary, and its order does not exceed n.

THEOREM: *Let $\alpha$ and $\beta$ be elementary functions of orders not exceeding n. Then $\alpha + \beta$, $\alpha - \beta$ and $\alpha\beta$ are elementary and their orders do not exceed n.*

We know that $\alpha$ and $\beta$ satisfy equations

$$\alpha^p + a_1 \alpha^{p-1} + \ldots + a_p = 0$$
$$\beta^q + b_1 \beta^{q-1} + \ldots + b_q = 0$$

with coefficients which are polynomials in monomials of orders $0, \ldots, n$. We let $k = pq$ and represent the k functions $\alpha^r \beta^s$, $r = 0, 1, \ldots, p-1$, $s = 0, 1, \ldots, q-1$, taken in any order, by $\xi_1, \ldots, \xi_k$. We consider any $\xi_i$, call it simply $\xi$, and write

$$\xi = \alpha^P \beta^Q, \quad 0 \leq P < p, \quad 0 \leq Q < q.$$

Consider $\alpha\xi$. If $P < p - 1$, $\alpha\xi$ is some $\xi_j$. If $P = p - 1$, we have

$$\alpha\xi = -(a_1 \alpha^{p-1} + \ldots + a_p) \beta^Q$$

so that

(8) $$\alpha\xi = c_1 \xi_1 + \ldots + c_k \xi_k$$

where the c are polynomials in monomials of orders $0, \ldots, n$. We have similarly, with d of obvious character,

(9) $$\beta\xi = d_1 \xi_1 + \ldots + d_k \xi_k.$$

Then
$$(\alpha + \beta) \mathcal{E} = (c_1 + d_1) \mathcal{E}_1 + \ldots + (c_k + d_k) \mathcal{E}_k.$$
Now unity is one of the $\mathcal{E}_i$ so that the $\mathcal{E}_i$ are not all zero. It follows from the lemma above that $\alpha + \beta$ is elementary and of order not greater than n. We proceed similarly for $\alpha - \beta$. To treat $\alpha\beta$, we use (8) and write
$$\alpha \beta \mathcal{E} = c_1 \beta \mathcal{E}_1 + \ldots + c_k \beta \mathcal{E}_k$$
so that, by (9), $\alpha\beta\mathcal{E}$ is linear in the $\mathcal{E}_i$.

It follows from the theorem just proved that if a function $\alpha$ is a a polynomial, with constant coefficients, in elementary functions of orders not exceeding n, then $\alpha$ is elementary and of order not more than n.

THEOREM: *Let an analytic function $\mu$ satisfy an equation*

(10) $$\mu^t + \alpha \mu^{t-1} + \beta \mu^{t-2} + \ldots + \lambda = 0$$

*where $\alpha, \beta, \ldots, \lambda$ are elementary and of orders not exceeding n. Then $\mu$ is elementary and its order is not more than n.*

Let $\alpha, \beta, \ldots, \lambda$ be given by equations with coefficients rational in monomials,
$$\alpha^p + a_1 \alpha^{p-1} + \ldots + a_p = 0,$$
$$\beta^q + b_1 \alpha^{q-1} + \ldots + b_q = 0,$$
$$\cdot \quad \cdot \quad \cdot \quad \cdot \quad \cdot \quad \cdot$$
$$\lambda^r + l_1 \lambda^{r-1} + \ldots + l_r = 0.$$

We let $k = pq \ldots rt$, and consider the k functions
$$\alpha^P \beta^Q \ldots \lambda^R \mu^T, \quad 0 \leq P < p, \ldots, 0 \leq T < t,$$
which functions are not all zero, since unity is one of them. We represent these functions, in some order, by $\mathcal{E}_1, \ldots, \mathcal{E}_k$. For each $\mathcal{E}$,
$$\mu \mathcal{E} = \alpha^P \beta^Q \ldots \lambda^R \mu^{T+1}.$$

Thus, if $T < t - 1$, $\mu \mathcal{E}$ is some $\mathcal{E}_i$. If $\mu = T - 1$, we use (10) and find that

(11) $$\mu \mathcal{E} = - \alpha^{P+1} \beta^Q \ldots \lambda^R \mu^T - \ldots - \alpha^P \beta^Q \ldots \lambda^{R+1}$$

Consider the first power product in the second member of (11). It is a $\mathcal{E}_i$ if $P < p - 1$. If $P = p - 1$, the usual reduction gives, for the power product, a linear combination of the $\mathcal{E}_i$. Thus, $\mu \mathcal{E}$ is a linear combination of the $\mathcal{E}_i$ with coefficients as desired, and our theorem is proved.

## LEMMAS ON ALGEBRAIC FUNCTIONS

**13.** Let $y = f(x_1, \ldots, x_r)$ be algebraic in its r variables and analytic at $(a_1, \ldots, a_r)$. Let $\varphi_1, \ldots, \varphi_s$, where $s < r$, be algebraic functions of $x_{s+1}, \ldots x_r$, analytic at $(a_{s+1}, \ldots, a_r)$, with

$$\varphi_i(a_{s+1}, \ldots, a_r) = a_i, \quad i = 1, \ldots, s.$$

When each $x_i$, $i = 1, \ldots, s$, is replaced in f by $\varphi_i$, f becomes a function $g(x_{s+1}, \ldots, x_s)$, analytic at $(a_{s+1}, \ldots, a_r)$. We wish to see that g is algebraic in all its variables.

Let the irreducible equation which defines f be

$$\alpha_0 y^p + \ldots + \alpha_p = 0.$$

The substitution of the several $\varphi$ for $x_1, \ldots, x_s$ may annul every $\alpha$. However, it is clearly legitimate to assume that $s = 1$ and to make the substitutions in succession. The replacement of $x_1$ by $\varphi_1$ cannot annul every $\alpha$ identically in $x_2, \ldots, x_r$. Otherwise the $\alpha$ would have a common factor. The algebraic character of g can then be shown by the methods of §12. In §14, we shall use the case of $s = 1$. In II, 4, we shall use an $s > 1$. The $\varphi_i$ will then be constants.

Taking another situation, let us suppose that f, given as above, is zero at $(a_1, \ldots, a_r)$, while $\partial f / \partial x_1$ is not zero there. The equation $f = 0$ defines $x_1$ as a function g of $x_2, \ldots, x_r$, analytic at $(a_2, \ldots, a_r)$, and equal to zero there. We wish to see that g is algebraic. In the equation defining f, the term $\alpha_p$ cannot be free of $x_1$. Otherwise, as $\alpha_p = 0$ when $f = 0$, we could not take $x_2, \ldots, x_r$ as independent variables and determine $x_1$ so as to make f zero. Thus g, which satisfies $\alpha_p = 0$, is algebraic.

## LIOUVILLE'S PRINCIPLE

**14.** The order of an elementary function is a perfectly definite integer. However, the structural scheme of a function of order $n > 0$, as presented in §11, is not at all unique. For instance, $e^{x^2}$ and $e^{-x} e^{x^2+x}$ are the same function of order 1.

Among all representations of a given function u of order n, there are certain ones which employ a least number of n-monomials; in any such representation, the $r_n$ in (N + 1) of §11 is not greater than the $r_n$ of any other representation of u. Representations which have this feature of economy will be employed throughout our work on integration, and we shall now establish for them an important principle.

Under the assumption that $r_n$ is a minimum, we shall show that *no algebraic relation can exist among the $r_n$ monomials of order n in u and monomials of order less than n*. To make this statement explicit, we refer to §11 and use the point a there mentioned. Let $\mathcal{E}_1, \ldots, \mathcal{E}_p$ be monomials of orders less than n. Suppose that a function

(12) $\qquad f(x_1^{(n)}, \ldots, x_{r_n}^{(n)}; \quad y_1, \ldots, y_p),$

algebraic in all its variables and analytic for $x_i^{(n)} = \theta_i^{(n)}(a)$, $y_i = \mathcal{E}_i(a)$, vanishes in the neighborhood of a when each $x_i^{(n)}$ is replaced by its $\theta_i^{(n)}$ and each $y_i$ by $\mathcal{E}_i$. We shall show that *f vanishes for all values of the $x^{(n)}$ close to the $\theta^{(n)}(a)$, if only each $y_i$ is replaced by $\mathcal{E}_i$*.

The principle which we have just formulated underlies all of Liouville's work on the elementary functions. We shall call it *Liouville's principle*. The proof amounts to nothing more than solving for one of the n-monomials in any nonidentical relation which may exist. When that monomial is replaced in u by the expression found for it, u acquires an expression involving too few n-monomials. A question connected with the implicit theorem accounts for the details which follow.

Suppose that (12) is not an identity in the $x^{(n)}$ for $y_j = \mathcal{E}_j$, $j = 1, \ldots, p$. We can find a point b, as close as one pleases to a, such that

(13) $\qquad f(x_1^{(n)}, \ldots, x_{r_n}^{(n)}; \mathcal{E}_1(b), \ldots, \mathcal{E}_p(b))$

does not vanish identically in the $x^{(n)}$.

Consider the partial derivatives of f, of all orders, cross-derivatives included, with respect to the $x^{(n)}$. Not all of them can vanish for $y_i = \mathcal{E}_i(b)$, $x_i^{(n)} = \theta_i^{(n)}(b)$. If they did, the function in (13) would be constant when each $x_i^{(n)}$ varies in the neighborhood of $\theta_i^{(n)}(b)$. The constant would be zero, since (13) vanishes for $x_i^{(n)} = \theta_i^{(n)}(b)$; this would contradict what we know of (13).

Working now in a neighborhood $\mathfrak{M}$ of $x = a$, let us suppose that all of the above-mentioned derivatives of f up to and including those of order $j^*$ vanish throughout $\mathfrak{M}$ when the variables in (12) are replaced by their monomials, but that some derivative of order $j + 1$ does not vanish at $x = b$, where b is in $\mathfrak{M}$. That the required j and b exist follows from what precedes. To fix our ideas, suppose that

$$G(x_1^{(n)}, \ldots, y_p)$$

---

* In this, we consider f to be its own derivative of order zero.

is a partial derivative which vanishes over $\mathfrak{M}$ but that the derivative of G with respect to $x_1^{(n)}$ does not vanish at b.

By §13, the equation G = 0 determines $x_1^{(n)}$ as an algebraic function of $x_2^{(n)}$ ..., $y_p$, analytic at $(\theta_2^{(n)}(b), ..., \xi_p(b))$ which reduces to $\theta_1^{(n)}$ for the familiar replacements. If we substitute this algebraic function for $x_1^{(n)}$ in (N + 1) of §11, and have regard to §13, we find a contradiction of the assumption that $r_n$ is a minimum.

One must bear well in mind that the y in (12) must be replaced by their monomials before we get a function which is identically zero. For instance, the function

$$F = (3 y_1 - y_2) x_4'',$$

which vanishes for every x when $y_1$ is replaced by $e^x$, $y_2$ by $e^x + \log 3$ and $x_1''$ by log log x, is certainly not zero identically in $y_1$, $y_2$ $x_1''$. It vanishes in x and $x_1''$ when $y_1$ and $y_2$ are replaced by their monomials.

*DIFFERENTIATION*

15. Let u be a function of order 1, of regular structure (§11) at some point $\underline{a}$. We wish to form du/dx for the neighborhood of $\underline{a}$. Let $f(x_1', ..., x_r'; x)$ be the algebraic function appearing in (II). The derivative of u can be written

$$\sum_{r=1} \frac{\partial f(\theta_1, ..., \theta_r; x)}{\partial x_i'} \frac{d\theta_i}{dx} + \frac{\partial f(\theta_1, ..., \theta_r; x)}{\partial x}$$

If $\theta_i$ is an exponential, $e^{v_i}$, its derivative is $\theta_i v_i'$. If $\theta_i$ is a logarithm, log $v_i$, its derivative is $v_i'/v_i$. In the logarithmic case, $v_i$ is not zero at $\underline{a}$, since $\theta_i$ is analytic at $\underline{a}$.

Let $\theta_1, ..., \theta_t$ be exponentials and the remaining $\theta$ logarithms. Let

$$g = \sum_{r=1}^{t} \frac{\partial f(x_1', ..., x_r'; x)}{\partial x_i'} x_i' v_i' + \sum_{i=t+1}^{r} \frac{\partial f}{\partial x_i'} \frac{v_i'}{v_i} + \frac{\partial f}{\partial x}.$$

Then g is algebraic in x and the x'. If we take $\rho_1$ of §11 sufficiently small, g will be analytic for $|x - a| < \rho_1, ..., |x_r' - \theta_r(a)| < \rho_1$. If now we take $\rho$ correspondingly small, so as to limit the variation of the monomials, we see that g reduces to the derivative of u for $|x - a| < \rho$ when each variable is replaced by its monomial.

It is now easy to treat a function u of any order n. Of the algebraic functions introduced in (I), ..., (N) of §11, there are some which are used for forming logarithmic monomials. As each monomial is analytic at $\underline{a}$, such an algebraic function is not zero when x = a and each accented x is its $\theta(a)$; the function is therefore distinct from zero if the x are close to these values. If now $\rho_1$ is taken sufficiently small and if $\rho$ is taken correspondingly small, so as to limit the variation of the monomials, we may assume that none of the algebraic functions which give logarithmic monomials vanish when x differs from a, and each accented x from its $\theta(a)$, by a quantity less than $\rho_1$ in modulus.

This understood, the formulas for the differentiation of composite functions show that *if u is a function of order n, of regular structure at a point a, there exists an algebraic function of the x, analytic for* $|x - a| < \rho_1, \ldots, |x_{r_n}^{(n)} - \theta_{r_n}^{(n)}| < \rho_1$ *which reduces to the derivative of u for* $|x - a| < \rho$ *when each variable is replaced by the monomial which corresponds to it.*

Thus if u is elementary and of order n, its derivative is elementary and of order not exceeding n.

**16.** The matter which we shall now examine takes care of itself quite well in Chapter II. Still it strikes one of the deeper notes of Liouville's theory and may well be lifted out from among other details.

Let u of order n be derived from

$$g(x_1^{\{n\}}, \ldots, x).$$

Suppose first that $\theta_1^{\{n\}}$ is an exponential, $e^v$. The algebraic function which yields $du/dx$ can be written

(14) $$\frac{\partial g}{\partial x_1^{\{n\}}} x_1^{\{n\}} \varphi + \text{other terms}.$$

In (14), $\varphi$ is an algebraic function of $x_1^{\{n-1\}}, \ldots, x$ which yields the derivative of v. The "other terms" are derivatives of g with respect to $x_2^{\{n\}}, \ldots, x$ times algebraic functions which yield the derivatives of $\theta_2^{\{n\}}, \ldots, x$.

If, in g, we replace $x_1^{\{n\}}$ by $\mu \theta_1^{\{n\}}$, where $\mu$ is a constant close to unity, and the other variables by the monomials which correspond to them, we secure a function of x, analytic at $\underline{a}$, which we shall call $u_\mu$. It is clear that $du_\mu/dx$ is obtained from (14) by replacing $x_1^{\{n\}}$ by $\mu \theta_1^{\{n\}}$ and the other variables by their monomials. If we consider $u_\mu$ as a function of x and $\mu$, it is seen to be analytic for $x = a$, $\mu = 1$.

Now let $\theta_1^{\{n\}}$ be a logarithm. We secure $du/dx$ from an algebraic function

(15) $$\frac{\partial g}{\partial x_1^{\{n\}}} \varphi + \text{other terms},$$

where $\varphi$, algebraic in $x_i^{(n-1)}, \ldots, x$, reduces to the derivative of $\theta_i^{(n)}$. Let $\mu$ be a constant close to zero and let $u_\mu$ be the function, analytic at $x = a$, which is obtained from g on replacing $x_i^{(n)}$ by $\theta_i^{(n)} + \mu$ and the other variables by their monomials. The same substitutions performed in (15) will produce the derivative of $u_\mu$. Furthermore $u_\mu$, as a function of x and $\mu$, is analytic at $x = a$, $\mu = 0$.

## SUMMARY

17. In §§ 3-16, we made a close examination of the elementary functions. In formal work, it suffices to bear a few facts in mind. In the structure of a function u of order n, there appear certain monomials of orders $0, 1, \ldots, n$. The function u is algebraic in these monomials, although some of the monomials may be used only for building monomials of higher order and may not appear effectively by themselves in the final expression for u. The derivative of u is algebraic in monomials which appear in the structure of u. If an expression for u is taken which employs as few as possible n-monomials, any algebraic relation among these n-monomials and any monomials of orders less than n, must hold identically in the monomials of order n. The reason is that a nonidentical relation would furnish an expression for one of the n-monomials which would permit us to write u with fewer such monomials. If $\theta$ is an n-exponential in u, and if $\theta$ is replaced in u by $\mu\theta$, where $\mu$ is a constant, we secure a function whose derivative is obtained from the derivative of u by the same substitution. If $\theta$ is an n-logarithm in u, and if $\theta$ is replaced in u by $\theta + \mu$, we secure a function whose derivative is obtained from that of u by the same substitution.

We are now prepared to take up questions of integration in finite terms.

## Chapter II

## ALGEBRAIC FUNCTIONS WITH ELEMENTARY INTEGRALS

### *LIOUVILLE'S FIRST THEOREM*

1. The meaning of the expression *integrable in finite terms* depends on the material with which one is working. An elementary function is said to be integrable in finite terms if its integral is elementary.

The first problem considered by Liouville in the field of integration in finite terms deals with the integration of algebraic functions.* We consider an algebraic function $y(x)$ defined by an irreducible relation

(1) $\quad\quad\quad\alpha_0(x) \, y^m + \alpha_1(x) \, y^{m-1} + \ldots + \alpha_m(x) = 0.$

In dealing with the integral of y, we shall need only the barest knowledge of the nature, as a function, of the integral. Any element $P(x - x_0)$ of y can be integrated into an element Q. The m.a.f. obtained from Q can be called *an integral of* y and can be written $\int y \, dx$. Q is determined to within an additive constant. By using all constants we obtain every integral of y, that is, every function whose elements have derivatives which are elements of y.

Most of the time we can work on an even simpler basis. Let us consider some branch of y which is analytic in some simply connected area $\mathfrak{U}$. Then $\int y \, dx$, which is defined to within an additive constant, is analytic throughout $\mathfrak{U}$. If now we ask whether y is integrable in finite terms, our question is a perfectly clear one. We are asking whether there is an elementary function $u(x)$, of regular structure** at a point $\underline{a}$ in $\mathfrak{U}$, whose derivative coincides with y for a neighborhood of $\underline{a}$. Any further questions which may arise can be taken care of by considerations of analytic continuation.

2. We now state Liouville's first theorem on integration.

THEOREM: *Let* $y(x)$ *be an algebraic function whose integral is elementary. Then*

(2) $\quad\quad\int y \, dx = v_0(x) + c_1 \log v_1(x) + \ldots + c_r \log v_r(x)$

*where r is a positive integer, each* $v(x)$ *an algebraic function, and each* c *a constant.*

---

\* Liouville [4]. A list of references is given at the end of the monograph.
\*\* I, 11 (Chapter I, §11). When no chapter is mentioned in a reference, the chapter is that in which one is reading.

Of course, the integral may be algebraic. This case may be considered to be covered by our statement if we allow all c to be zero.

An expression like the second member of (2) has an algebraic derivative. As one approaches the problem, it appears to be a foregone conclusion that the only possibility for an elementary integral is that described in (2). If the answer contained an exponential, the exponential ought to survive after differentiation. Similarly one cannot imagine a logarithm disappearing unless it enters linearly. Considerations of this vague type had led Laplace, before Liouville's time, to conjecture the theorem established by Liouville, and Liouville claimed for his method of proof the merit of following these intuitive ideas.

The proof of Liouville's theorem will be conducted as follows. Letting u be an integral of y, we shall suppose that u is elementary but not algebraic. Let n > 0 be the order of u. We shall suppose that we have for u an expression of the type described in I, 14, which involves as few as possible n-monomials, as in I, 11. We shall show first that none of the n-monomials is an exponential. It will then be shown that u has an expression like the second member of (2) with each $\log v_i$ an n-logarithm and $v_0$ a function of order less than n. The proof will be completed by showing that $n = 1$.

3. As has just been stated, we understand $r_n$ to be small as possible. We work at a point $\underline{a}$ at which u has regular structure. Let the algebraic function of (N + 1) of I, 11, from which u is obtained be represented by

$$g(x_1^{(n)}, \ldots, x).$$

Let every variable except $x_1^{(n)}$ be replaced by its monomial. We obtain a function of $x_1^{(n)}$ and x which we shall represent by $f(x_1^{(n)}, x)$. This latter function produces u when $x_1^{(n)}$ is replaced by $\theta_1^{(n)}$. On this basis, letting $\theta$ represent $\theta_1^{(n)}$, we write

(3) $$u = f(\theta, x).$$

We now obtain the derivative of u, having regard to I, 15. On the basis of (3), we write

(4) $$\frac{du}{dx} = f_\theta(\theta, x) \frac{d\theta}{dx} + f_x(\theta, x).$$

The expression $f_\theta$ is obtained by replacing $x_1^{(n)}$ by $\theta$ in $\partial f(x_1^{(n)}, x)/\partial x_1^{(n)}$. The description of $f_x$ is similar. We notice also that the second member of (4) may be considered to be obtained by replacements from (14) or (15) of Chapter I.

Let us suppose now that $\theta$ is an exponential and let $\theta = e^w$ with $w$ of order $n-1$. Observing that the derivative of $u$ is $y$, we obtain from (4)

$$(5) \qquad y = f_\theta(\theta, x) \; \theta \; \frac{dw}{dx} + f_x(\theta, x).$$

Now (5) is an algebraic relation among the monomials in $u$, of the type studied in I, 14. By Liouville's principle (5) must hold identically in $\theta$. This means, on the one hand, that if, in the algebraic function out of which the second member of (5) is obtained, we replace all variables except $x_i^{(n)}$ by their monomials, we will get an expression in $x_i^{(n)}$ and $x$ which is identical with $y(x)$ for every $x_i^{(n)}$. It means again, perhaps more simply, that the relation

$$y = f_{x_i^{(n)}}(x_i^{(n)}, x) \; x_i^{(n)} \; \frac{dw}{dx} + f_x(x_i^{(n)}, x)$$

is an identity in $x_i^{(n)}$ and $x$.

In particular, we may replace $\theta$ in (5) by any function of $x$ which is analytic at the point $\underline{a}$ and close in value to $\theta(a)$ at $\underline{a}$. We shall replace $\theta$ by $\mu\theta$ where $\mu$ is a constant close to unity. We may thus write, for $x$ close to $\underline{a}$,

$$(6) \qquad y = f_{\mu\theta}(\mu\theta, x) \; \mu\theta \; \frac{dw}{dx} + f_x(\mu\theta, x)$$

where $f_{\mu\theta}$ is the result of replacing $x_i^{(n)}$ by $\mu\theta$ in $f_{x_i^{(n)}}$.

By 1, 15, and even on the basis of its own structure, the second member of (6) is the derivative of $f(\mu\theta, x)$. By (5) and (6), $f(\theta, x)$ and $f(\mu\theta, x)$ have the same derivative and thus differ by a constant. This constant depends on $\mu$. We write

$$(7) \qquad f(\mu\theta, x) = f(\theta, x) + \beta(\mu).$$

As $f(\mu\theta, x)$ is analytic in $\mu$ and $x$ for $\mu = 1$, $x = a$, the function $\beta(\mu)$ must be analytic for $\mu = 1$.

We differentiate (7) with respect to $\mu$ and find

$$(8) \qquad f_{\mu\theta}(\mu\theta, x) \; \theta = \beta'(\mu),$$

the accent indicating differentiation. We put $\mu = 1$ and represent $\beta'(1)$ by $c$, obtaining the relation

$$(9) \qquad f_\theta(\theta, x) \; \theta = c.$$

Again we apply Liouville's principle; (9) holds identically in $\theta$. We replace $\theta$ by an independent variable $z$ and have, for the neighborhood of $x = a$, $z = \theta(a)$,

$$(10) \qquad z \; f_z(z, x) = c.$$

ALGEBRAIC INTEGRANDS

Now (10) is a partial differential equation for $f(z, x)$. It gives
$$f_z(z, x) = \frac{c}{z}$$
so that

(11) $\qquad f(z, x) = c \log z + \gamma(x),$

where $\gamma$ is analytic at $x = a$. We notice that, as $\theta(a) \neq 0$, $\log z$ is analytic for $z = \theta(a)$.

To determine $\gamma(x)$, we replace $z$ in (11) by any value $z_0$ close to $\theta(a)$. We have
$$f(z_0, x) = c \log z_0 + \gamma(x).$$
Hence, identically in $z$ and $x$,

(12) $\qquad f(z, x) = c \log z - c \log z_0 + f(z_0, x).$

In particular, we may replace $z$ by $\theta$ in (12) for the neighborhood of $x = a$. Thus

(13) $\qquad u = f(\theta, x) = c \log \theta - c \log z_0 + f(z_0, x).$

In (13), $\log \theta$ is of order $n-1$ while $f(z_0, x)$, which by I, 13, is an elementary function of regular structure at $\underline{a}$, involves fewer than $r_n$ monomials of order n. This contradicts the fact that the expression (3) of u is as economical as possible in n-monomials. We have thus proved that $\theta$ is not an exponential.

4. We know now that the $\theta_i^{(n)}$ are logarithms. We shall prove that u is a function of order less than n plus terms of the form $c_i \theta_i^{(n)}$ with constant c. We use (4). Let $\theta = \log w$ with w of order $n-1$. Then

(14) $\qquad y = f_\theta(\theta, x) \frac{w'}{w} + f_x(\theta, x).$

As w and w' are of order less than n, (14) is an identity in $\theta$.

We replace $\theta$ by $\theta + \mu$ where $\mu$ is a constant close to zero. Then

(15) $\qquad y = f_{\theta+\mu}(\theta+\mu, x) \frac{w'}{w} + f_x(\theta+\mu, x).$

The second member of (15) is the derivative of $f(\theta+\mu, x)$. Then

(16) $\qquad f(\theta+\mu, x) = f(\theta, x) + \beta(\mu)$

where $\beta$ is analytic for $\mu = 0$. Differentiation of (16) with respect to $\mu$ gives
$$f_{\theta+\mu}(\theta+\mu, x) = \beta'(\mu).$$
Putting $\mu = 0$ and writing $\beta'(0) = c_1$, we have

(17) $\qquad f_\theta(\theta, x) = c_1.$

Now (17) is an identity in $\theta$. We replace $\theta$ by a variable $z$ and write
$$f_z(z, x) = c_1.$$
Then
$$f(z, x) = c_1 z + \gamma(x).$$
Determining $\gamma$ by a special value $z_0$ of $z$, we have
$$f(z, x) = c_1 z - c_1 z_0 + f(z_0, x).$$
Hence
(18) $\qquad u = f(\theta, x) = c_1 \theta - c_1 z_0 + f(z_0, x).$

In $g(x_i^{(n)}, \ldots, x)$ of §3, let the variables corresponding to monomials of order less than $n$ be replaced by their monomials. There results a function
$$h(x_1^{(n)}, \ldots, x_r^{(n)}; x)$$
where $r$ represents $r_n$. We have
(19) $\qquad u = h(\theta_1^{(n)}, \ldots, \theta_r^{(n)}; x).$

We equate the second members of (18) and (19), remembering that $\theta = \theta_1^{(n)}$. The resulting equation holds identically in the $\theta^{(n)}$. This means that
$$\frac{\partial h}{\partial x_1^{(n)}} = c_1.$$
There exist similarly, for $i = 2, \ldots, r$, constants $c_i$ such that
$$\frac{\partial h}{\partial x_i^{(n)}} = c_i.$$
Thus
(20) $\qquad h = c_1 x_1^{(n)} + \ldots + c_r x_r^{(n)} + v(x)$
with $v(x)$ analytic for $x = a$. To determine $v(x)$, we replace the $x^{(n)}$ in (20) by constants. We find, referring to I, 13, that $v(x)$ is elementary, of regular structure at $\underline{a}$ and of order less than $n$. We have, by (20),
$$u = v(x) + c_1 \theta_1^{(n)} + \ldots + c_r \theta_r^{(n)}$$
as was to be proved.

5. We now prove that $n = 1$. Let us assume that $n > 1$. We know that $u$ has an expression
(21) $\qquad c_1 \log v_1 + \ldots + c_r \log v_r + v$
with each $\log v_i$ of order $n$ and $v$ of order less than $n$.* There will

* It should be emphasized that $r$ is the least number of $n$-monomials in terms of which $u$ can be expressed.

be many expressions (21) for u. These will all have the same r, but different $v_1$, v, and $c_1$. In each expression, a certain number of (n-1)-monomials are used in building the r + 1 functions $v_1$, ..., $v_r$, v found in that expression. We consider those expressions which use as few as possible (n-1)-monomials.

From the expressions just described, we select one in which as few as possible (n-1)-monomials are used in building the r functions $v_1$, ..., $v_r$. We understand (21) to be the expression just selected. Let θ be one of the (n-1)-monomials appearing in the $v_i$. We write

(22)  $u = c_1 \log v_1(\theta, x) + \ldots + c_r \log v_r(\theta, x) + v(\theta, x).$

We have

(23)  $y = \sum_{j=1}^{r} c_j \frac{v_{j\theta}(\theta, x) \theta' + v_{jx}(\theta, x)}{v_j(\theta, x)} + v_\theta(\theta, x) \theta' + v_x(\theta, x).$

where literal subscripts indicate partial differentiation.

We show first that θ cannot be an exponential. Suppose that $\theta = e^w$ with w of order n-2. Then

(24)  $y = \sum_{j=1}^{r} c_j \frac{v_{j\theta}(\theta, x) \theta w' + v_{jx}(\theta, x)}{v_j(\theta, x)} + v_\theta(\theta, x) \theta w' + v_x(\theta, x).$

In (24) we have an algebraic relation among the (n-1)-monomials in u and monomials of order less than n-1. The relation (24) must be an identity in θ. If it were not, we could, with all the formality of I, 14, solve for θ in (24) and, by a substitution, express u with fewer (n-1)-monomials than are found in (22).

We replace θ by μθ in (24). The second member becomes the derivative of

$\sum_{j=1}^{r} c_j \log v_j(\mu\theta, x) + v(\mu\theta, x).$

Hence

$\sum_{j=1}^{r} c_j \log v_j(\mu\theta, x) + v(\mu\theta, x) = u(x) + \beta(\mu)$

for every μ close to unity and for every x close to the point a at which our functions are being studied. We differentiate with respect to μ and put μ = 1. Then

(25)  $\sum_{j=1}^{r} c_j \frac{v_{j\theta}(\theta, x) \theta}{v_j(\theta, x)} + v_\theta(\theta, x) \theta = \beta'(1) = c.$

Again, (25) is an identity in θ. Thus, for a variable z,

$\sum_{j=1}^{r} c_j \frac{v_{jz}(z, x)}{v_j(z, x)} + v_z(z, x) = \frac{c}{z}.$

Then

$$\sum_{j=1}^{r} c_j \log v_j(z, x) + v(z, x) = c \log z + \gamma(x)$$
$$= c \log z - c \log z_0 + \sum_{j=1}^{r} \log v_j(z_0, x) + v(z_0, x)$$

where $z_0$ is close to $\theta(a)$. For $z = \theta(x)$ we find

(26) $\quad u = c \log \theta - c \log z_0 + \sum_{j=1}^{r} c_j \log v_j(z_0, x) + v(z_0, x).$

In the second member of (26), each $\log v_j$ must be of order n; otherwise u would either be of order less than n or be expressible with less than r n-monomials. Now $\log \theta = w$, so that (26) gives an expression of type (21) for u involving fewer (n-1)-monomials than appear in the expression selected for u. Thus, $\theta$ cannot be an exponential.

We now assume that $\theta$ is a logarithm, Let $\theta = \log w$. Then (23) gives

(27) $\quad y = \sum \frac{c_j}{v_j(\theta, x)} [v_{j\theta}(\theta, x) \frac{w'}{w} + v_{jx}(\theta, x)] + v_\theta(\theta, x) \frac{w'}{w} + v_x(\theta, x),$

an equation which must hold identically in $\theta$. We replace $\theta$ by $\theta + \mu$ and find that

(28) $\quad \Sigma c_j \log v_j(\theta+\mu, x) + v(\theta+\mu, x) = u(x) + \beta(\mu)$

with $\beta(\mu)$ analytic for $\mu = 0$. We differentiate (28) with respect to $\mu$ and put $\mu = 0$. Then

$$\Sigma c_j \frac{v_{j\theta}(\theta, x)}{v_j(\theta, x)} + v_\theta(\theta, x) = \beta'(0) = c.$$

As this holds identically in $\theta$, we write

$$\Sigma c_j \frac{v_{jz}(z, x)}{v_j(z, x)} + v_z(z, x) = c.$$

Integrating, we have

$$\Sigma c_j \log v_j(z, x) + v(z, x) = cz + \gamma(x)$$
$$= cz - cz_0 + \Sigma c_j \log v_j(z_0, x) + v(z_0, x).$$

For $z = \theta(x)$,

(29) $\quad u = c\theta - cz_0 + \Sigma c_j \log v_j(z_0, x) + v(z_0, x).$

In (29), we have an expression in which the (n-1)- monomials are among those in (22). The n-logarithms in (29) employ fewer (n-1)-monomials than appear in (22). This final contradiction renders untenable the assumption that n > 1, and the proof of Liouville's theorem is completed.

6. Let us try to describe Liouville's method of proof, as represented, in particular, by the work of §§ 3 and 4. We have a function u of order n whose structure is to be examined. We use an expression for u which involves a least number of n-monomials. Let θ be one of these. When we differentiate, we secure a relation which holds identically in θ. We replace θ by μθ if θ is an exponential and by θ + μ if θ is a logarithm. We find that when one of these substitutions is made in u, that function is increased by a function of μ. The relation thus secured is differentiated with respect to μ; μ is put equal to unity in the exponential case and to zero in the logarithmic case. We apply Liouville's principle again, replacing θ by a variable z. We have now a partial differential equation, which we integrate, determining the arbitrary function of x which appears by using a special value of z. Replacing z by θ, we have an expression for u which permits us to draw conclusions as to the structure of that function.

In what way did Liouville, working on these questions at the age of about twenty-three, assemble the ideas which underlie his method? It is a simple theory that he found inspiration in Abel's investigation of the unsolvability of the quintic equation.* Abel gives a classification of expressions involving radicals which resembles Liouville's arrangement of the elementary functions. He selects, for a given function, the expression most economical in radicals and is then able to replace certain radicals by arbitrary quantities.

7. Let us take another look at Liouville's procedure. If the integral u is elementary and if proper economy in monomials is observed, each exponential in u may be replaced by a constant times itself and each logarithm by a constant plus itself; the replacements will leave u an integral. Now the integrals of y are a one-parameter family. The monomials which enter into u must not create two or more arbitrary constants. Here we have a viewpoint from which one can conjecture Liouville's theorem and the answers to other questions on integration in finite terms. This idea is due to Koenigsberger [2]. It underlies the proofs in Mordukhai-Boltovskoi's paper [15].

---

* Abel [1]. The criticisms which have been made of Abel's reasoning appear to be unfounded.

## ABEL'S THEOREM

8. It was shown by Abel, about a decade before Liouville proved the theorem of §2, that when the integral of an algebraic function $y(x)$ is expressible as in (2), the algebraic functions $v$ may be taken so as to be rational in $x$ and $y$, with constant coefficients. In the language of the theory of algebraic functions, we may arrange so that the $v$ are rational on the Riemann surface of $y$.

To secure a perspicuous proof of Abel's theorem, we shall permit ourselves to use a few of the simpler qualitative properties of algebraic functions.*

As we understand the relation (2), on the basis of our derivation of it, $y$ and the $v$ are analytic in some definite area $\mathfrak{U}$. Now $y$ is defined by (1) and the $v_i$ by similar equations. We have been working with definite branches of $y$ and the $v$. We shall suppose, shrinking $\mathfrak{U}$ if necessary, that every branch of $y$, and every branch of each $v$, is analytic in $\mathfrak{U}$. We shall suppose that (2) has been established for branches

(30) $\qquad\qquad y_1, v_{01}, \ldots, v_{r_1}$

of $y$ and the $v$.

Let $\mathfrak{a}$ be a point in $\mathfrak{U}$ and let $C$ be a curve which starts and ends at $\mathfrak{a}$, avoiding the singular points of $y$ and the $v$. When we continue $y_1$ and the $v_{i1}$ along $C$, the set (30) is replaced by some set of branches, perhaps (30) itself, of $y$ and the $v$. We shall employ only curves $C$ which replace $y_1$ by itself. The number of such curves is infinite, but there are only a finite number of possibilities for the sets by which (30) can be replaced. Let the distinct sets which it is possible to secure be

(31)
$$\begin{array}{c} y_1, v_{01}, \ldots, v_{r_1} \\ y_1, v_{02}, \ldots, v_{r_2} \\ \cdot \quad \cdot \quad \cdot \quad \cdot \quad \cdot \\ y_1, v_{0t}, \ldots, v_{rt} \end{array}$$

For instance, in the second column of (31), $v_{01}, \ldots, v_{0t}$ are various branches of $v_0$. They need not be distinct branches; it is only the sets as a whole which are distinct. Let $C_i$ be a curve which converts the first set into $y_1, v_{0i}, \ldots, v_{ri}$.

Now let

---

* In Chapter III a more general question will be treated by algebraic methods.

(32) $$w_0 = v_{01} + v_{02} + \ldots + v_{0t}.$$

We are going to prove that $w_0$ is unchanged when it is continued along a curve C, starting and ending at <u>a</u>, which returns $y_1$ to itself. For this we prove that C permutes the sets of (31) among themselves. To continue the $i^{th}$ set along C is to continue the first set along $C_i$ and then along C. Thus every set goes into some set when it is continued along C. Two distinct sets cannot go into the same set, since the continuation along C is a reversible process.

On this basis, when $w_0$ is continued along C, the second subscripts of the $v_{0i}$ in (32) are permuted among themselves. This shows that $w_0$ is unchanged.

By I, 12, $w_0$ is a branch of an algebraic function of x. We are going to show that this algebraic function is rational in y and x. Consider a curve $D_1$ which starts at <u>a</u> and ends at a point b, avoiding singular points of y and $v_0$. Along it, $y_1$ and $w_0$ can be continued. Let $P(x - b)$ be the element secured for $y_1$ at b and $Q(x - b)$ the element secured for $w_0$. If a second path $D_2$ continues $y_1$ from <u>a</u> into $P(x - b)$, then $D_2$ must continue $w_0$ into $Q(x - b)$. Otherwise, using $D_1$, and $D_2$ reversed, we would have a path which leaves $y_1$ unchanged while changing $w_0$. Thus the continuation of $w_0$ furnishes an algebraic function which is one-valued on the Riemann surface of y and is therefore a rational combination of y and x.

Let us explain the point just made in greater detail. Let the branches of y, analytic in $\mathfrak{U}$, be $y_1, \ldots, y_m$. We denote $w_0$ by $w_{01}$. All paths which continue $y_1$ into $y_i$ continue $w_{01}$, as was seen above, into a single definite function analytic in $\mathfrak{U}$, which we denote by $w_{0i}$. Now let functions A be determined by equations

$$w_{01} = A_0 + A_1 y_1 + \ldots + A_{m-1} y_1^{m-1}$$
$$w_{02} = A_0 + A_1 y_2 + \ldots + A_{m-1} y_2^{m-1}$$
$$\cdot \quad \cdot \quad \cdot \quad \cdot \quad \cdot \quad \cdot \quad \cdot \quad \cdot$$
$$w_{0m} = A_0 + A_1 y_m + \quad + A_{m-1} y_m^{m-1}.$$

The determinant D of this system is not zero, for it is a Vandermonde determinant and the $y_i$ are distinct. we have

$$A_j = D^{-1} D_j$$

where $D_j$ is a determinant with the $w_{0i}$ in one column. By I, 12, $A_j$ is an algebraic function. We wish to see that $A_j$ is unchanged when continued along any path which starts and ends at <u>a</u>. For this we notice that such a continuation permutes the rows of D among themselves and performs the same permutation on $D_j$. Thus the continuation either leaves D and $D_j$ unchanged or replaces them both by their negatives, so that $A_j$ is unchanged. An algebraic function which is unchanged in this manner can have but one branch and so is rational. To sum up,

$$w_0 = A_0 + A_1 y + \ldots + A_{m-1} y^{m-1}$$

with each A a rational function of x.

For $i > 0$, we use a product rather than a sum. If $w_i = v_{i1} v_{i2} \ldots v_{it}$, $i = 1, \ldots, r$, $w_i$ is a rational combination of y and x.

We return now to (2). Let us show that, for $i = 1, \ldots, t$, we have a relation

(33) $\qquad d_i + \int y \, dx = v_{0i} + c_1 \log v_{1i} + \ldots + c_r \log v_{ri}$

where the d are constants. We work in the neighborhood of a point $\underline{a}$ in $\mathfrak{U}$ at which no $v_{ji}$ with $j > 0$ is zero. Each log $v_{ji}$ will have an element at $\underline{a}$, secured in a definite manner.

For $i = 1$, $d_1 = 0$ and (33) is merely (2). For any other i, we use a curve $C_i$ which continues the first set of (31) into $y_1$, $v_{01}$, ..., $v_{r1}$, avoiding singularities and values of x at which one or more $v_{j1}$ have zero values. Then each log $v_{j1}$ is continued into a definite function log $v_{ji}$. Since $y_1$ returns to itself, its integral is changed through the addition of a constant, which we call $d_i$.

Summing the equations (33) for $i = 1, \ldots, t$, we find

(34) $\qquad h + t \int y \, dx = w_0 + c_1 \log w_1 + \ldots + c_r \log w_r$

where h is the sum of the d. Thus

(35) $\qquad \int y \, dx = \dfrac{w_0 - h}{t} + \dfrac{c_1}{t} \log w_1 + \ldots + \dfrac{c_r}{t} \log w_r.$

In the second member of (35), the algebraic functions are all rational in y and x. Equation (35) has been established for the neighborhood of $\underline{a}$. It is preserved under analytic continuation. Perhaps the simplest way to look at the situation is as follows. In (2), y is a function with a definite Riemann surface. It has been shown that the v may be selected so as to have the same surface. Each logarithmic derivative $v'_i/v_i$ is a function rational on this surface, and we have

(36) $\qquad y = v'_0 + c_1 \dfrac{v'_1}{v_1} + \ldots + c_r \dfrac{v'_r}{v_r}.$

Every idea contained in (2) is contained in the simpler relation (36).

## ALGEBRAIC FUNCTIONS WITH ALGEBRAIC INTEGRALS

9. As a special case of Abel's theorem, we have the result that *if y is an algebraic function of x and if the integral of y is algebraic, the integral is rational in y and x.*

For this special case we shall present a separate proof, almost entirely algebraic.

Let u, the integral, be analytic in $\mathfrak{A}$. Let the irreducible equation satisfied by u be

(37) $$B_0 u^p + \ldots + B_p = 0,$$

the B being polynomials in x with $B_0 \neq 0$. The existence of (37) is enough to show that u satisfies in $\mathfrak{A}$ various equations of the form

(38) $$D_0 u^q + \ldots + D_q = 0$$

with each D a polynomial in y and x with constant coefficients, $D_0$ not vanishing identically in x in $\mathfrak{A}$. We have in (37) a trivial instance of (38). From among all equations of type (38) satisfied by u, we select one which is of a least degree in u. We shall suppose that (38) is such an equation of least degree and proceed to prove that $q = 1$.

Suppose that $q > 1$. We write (38)

(39) $$u^q + \beta_1(x) u^{q-1} + \ldots + \beta_q(x) = 0$$

with each $\beta$ rational in y and x. Differentiating (39), we have

(40) $$u^{q-1} [q y + \frac{d\beta_1}{dx}] + u^{q-2} [(q-1) y \beta_1 + \frac{d\beta_2}{dx}] + \ldots + [y \beta_{q-1} + \frac{d\beta_q}{dx}] = 0.$$

Now

$$\frac{d\beta_1}{dx} = \frac{\partial \beta_1}{\partial x} + \frac{\partial \beta_1}{\partial y} \frac{dy}{dx}$$

The derivative of y is found from (1) to be rational in y and x. Thus the coefficients in (40) are rational in y and x. If those coefficients were not zero identically in x, we would, clearing fractions, have an equation like (38) for u, of degree less than q. Thus, throughout $\mathfrak{A}$,

$$q y + \frac{d\beta_1}{dx} = 0$$

and

(41) $$u = \int y\, dx = \frac{1}{q}\beta_1(x, y) + c.$$

Now (41) is an equation (38) for u with q = 1. We have thus a contradiction of the assumption that q > 1. Then q = 1 and (38) expresses u rationally in y and x.

As to the nature of the proof, the selection of an equation (38) of least degree is a device which replaces a reduction algorithm. We could instead start by differentiating (37) after dividing by $B_0$ and replace dy/dx by its expression in terms of y and x. We would either secure an equation (38) of degree less than p or else obtain an expression for the integral as in (41). In the former case, the process would be repeated until a rational expression for u is secured.

From the standpoint of the theory of abelian integrals, the result on algebraic integrals is a very obvious one. If the integral of y is algebraic, it can have no periods, either cyclic or polar. It is thus uniform on the surface of y and so is rational in y and x.

10. Liouville [3] furnished a method for determining whether an algebraic function has an algebraic integral and for obtaining the integral when it is algebraic. Without following the matter through to the very end, let us see what is involved in it.

If $y(x)$, defined by (1), has an algebraic integral, we have, by §8,

$$\int y\, dx = A_0 + A_1 y + \ldots + A_{m-1} y^{m-1}$$

with each A a rational function of x. Then

(42) $$y = \frac{dA_0}{dx} + \sum_{i=1}^{m-1} (y^i \frac{dA_i}{dx} + i A_i y^{i-1} \frac{dy}{dx}).$$

By (1) each $y^{i-1}$ dy/dx is a rational combination of x and y. Every such rational combination is a linear combination of 1, y, ..., $y^{m-1}$ with rational functions of x for coefficients. Let such linear combinations be substituted for each $y^{i-1}$ dy/dx in (42). Then (42) becomes

(43) $$y = \sum_{i=0}^{m-1} y^i (\frac{dA_i}{dx} + \beta_i)$$

with each β a linear combination of $A_1$, ..., $A_{m-1}$ with rational functions of x for coefficients. The expression of a function as a linear combination of 1, ..., $y^{m-1}$ is unique; two such expressions for a single function would furnish an equation for y of degree less than m. Thus we may equate coefficients of like powers of y in both members of (43). This furnishes for the A a system of linear differential equations of the first order, with known rational functions of x for coefficients. Our problem is to determine whether this system has a solution with each A rational. Liouville carries this question through. We shall not go into the details since, from the standpoint of the theory of linear differential equations as it exists at present, the question is an essentially routine one. In III, 11 we work out a simple problem of this type.

No general method exists for determining whether an algebraic function whose integral is not algebraic can be integrated with logarithms. This problem, which depends on delicate questions in the theory of numbers, appears to be a difficult one.

## RESIDUES AND INTEGRATION

11. The methods which were at Liouville's disposal for the examination of special algebraic functions were of a somewhat laborious type. Great simplicity can be secured by the use of the expansions of an algebraic function at places on its Riemann surface. We shall recall the facts.

Let $y(x)$ be algebraic and let $u(x)$ be an algebraic function, not identically zero, which is uniform on the surface of y; u thus is rational in x and y. At a point on the surface of y which is neither a branch point nor a point at $\infty$, u either is analytic or has a pole; it will have a Taylor development at the point in the first case and a Laurent development in the second. Now consider a branch point, corresponding to a finite value $x_0$ of x. There u has an expansion

$$(44) \qquad a_0(x - x_0)^{p/r} + a_1(x - x_0)^{(p+1)/r} + \ldots$$

where p is some integer and r the number of sheets which circulate at the point. We suppose that $a_0 \neq 0$. If $p > 0$, u is said to have a zero of order p at the point, and if $p < 0$, u is said to have there a pole of order $-p$. At each point on the surface at $\infty$, u has an expansion

$$(45) \qquad a_0 x^{p/r} + a_1 x^{(p-1)/r} + \ldots$$

with $a_0 \neq 0$. If the point is not a branch point, $r = 1$. There is a pole of order p if $p > 0$ and a zero of order $-p$ if $p < 0$.

Consider a finite point on the surface of y at which u has a pole. We write the expansion of u at this point in the form (44), with

$r = 1$ if the point is not a branch point. The product by r of the coefficient of $(x - x_0)^{-1}$ in (44) is called the *residue* of u at the pole. At a point at $\infty$, we call the product by $-r$ of the coefficient of $x^{-1}$ in the expansion (45) of u, the residue of u at the point. Thus there are two types of places at which we speak of residues, the poles at finite points and all points at $\infty$.

We now consider $u'/u$, the logarithmic derivative of u, which is uniform on the surface of y. We examine a finite point on the surface, at which u has an expansion (44). If $p = 0$, the expansion of $u'/u$, found formally from (44), will contain no negative powers if $r = 1$. If $r > 1$, the expansion may contain negative powers, but the exponents of $x - x_0$ will all exceed $-1$. Thus $u'/u$ may have a pole at the point if $r > 1$, but, if so, the residue is zero. If $p \neq 0$, the expansion of $u'/u$ starts with

$$r^{-1} p(x - x_0)^{-1}$$

so that $u'/u$ has a pole with p for residue. For a point at $\infty$, we find from (45) that the residue of $u'/u$ is $-p$.

Consider now $u'$. None of its expansions can contain a term of exponent $-1$. Thus the residues of $u'$ are all zero.

12. We return now to Liouville's theorem of §2, supposing that the integral of y is elementary but not algebraic. A set of r complex numbers $c_1, \ldots, c_r$ will be called *independent* if there does not exist a set of rational numbers $q_1, \ldots, q_r$, not all zero, such that

$$q_1 c_1 + \ldots + q_r c_r = 0.$$

Let an expression (2) of the integral of y be given which contains a minimum number of logarithms. We shall prove that $c_1, \ldots, c_r$ are independent. Suppose, for instance, that

$$c_1 = s_2 c_2 + \ldots + s_r c_r$$

with rational s. We can write the second member of (2) as

$$v_0 + c_2 \log v_1^{s_2} v_2 + \ldots + c_r \log v_1^{s_r} v_r.$$

In this expression, the functions of which logarithms are taken are algebraic. Thus r is not a minimum. This proves the independence of the c.

When we have an expression for $\int y\, dx$ which contains a least number of logarithms, we can, as in §8, replace it by an expression with the same number of logarithms and with each algebraic function rational in y and x.

We shall use, for $\int y\, dx$, such an expression as has just been described. We refer to (2). No $v_i$ with $i > 0$ is a constant. An algebraic function which is not a constant has at least one zero and at least one pole. Suppose, for instance, that $v_1$ has a zero at a finite point on the surface of y, for which $x = x_0$. At this point $v_1'/v_1$ will have a residue distinct from zero. Let the coefficient of $(x - x_0)^{-1}$ in the expansion of $v_i'/v_i$, at the point under consideration be $q_i$, $i = 1, \ldots, r$. The q are all rational and $q_1 \neq 0$. As the residues of $v_0'$ are all zero, we see from (2) that $q_1 c_1 + \ldots + q_r c_r$ is not zero so that y has a nonzero residue at the point. Similarly, if $v_1$ has a zero at a point at $\infty$, y has a nonzero residue at the point.

Thus, *if $y(x)$ is an algebraic function whose integral is elementary but not algebraic, there must exist points on the surface of y at which y has residues distinct from zero.*

## ELLIPTIC INTEGRALS

13. We shall now demonstrate the nonelementary character of Legendre's elliptic integrals of the first and second kinds,

$$I_1 = \int \frac{dx}{\sqrt{(1-x^2)(1-k^2 x^2)}}, \quad I_2 = \int \frac{x^2\, dx}{\sqrt{(1-x^2)(1-k^2 x^2)}},$$

where k is a constant with $k^2 \neq 0, 1$. We must show first that there are no nonzero residues and then that the integrals are not algebraic.

For those versed in the theory of abelian integrals, the nonelementary character of $I_1$ follows immediately from §12. An integral of the first kind, on a surface of any positive genus, has an infinite number of branches and has no logarithmic branch points. Thus no integral of the first kind can be elementary. An integral of the second kind cannot be elementary if it is not algebraic. We prefer, however, to treat the question by more elementary methods.

Consider $I_1$. Let
$$y = \sqrt{(1-x^2)(1-k^2 x^2)}.$$
The poles of $1/y$ are at $\pm 1, \pm k$. Writing

(46) $$\frac{1}{y} = \frac{(x-1)^{-1/2}}{\sqrt{(-x-1)(1-k^2 x^2)}},$$

we see that the function under the radical in (46) is not zero at $x = 1$, so that the reciprocal of its square root is analytic for

$x = 1$. Thus the residue of $1/y$ at $x = 1$ is zero. The same is true at the other three branch points. At $\infty$, the two branches of $1/y$ are uniform, and their developments start with $\pm k^{-1} x^{-2}$. Thus the residues at $\infty$ are zero.*

It remains to be shown that $I_1$ is not algebraic. Suppose that it is. By §§ 8 and 9

(47) $$I_1 = A_0 + A_1 y$$

where $A_0$ and $A_1$ are rational functions of $x$. We differentiate (47). Then

$$\frac{1}{y} = \frac{dA_0}{dx} + y\frac{dA_1}{dx} + A_1 \frac{dy}{dx} = \frac{dA_0}{dx} + y\frac{dA_1}{dx} + \frac{A_1}{2y}\frac{dy^2}{dx}.$$

Thus

(48) $$y\frac{dA_0}{dx} = 1 - y^2 \frac{dA_1}{dx} - \frac{A_1}{2}\frac{dy^2}{dx}.$$

The second member of (48) is a rational function of $x$. The first member is irrational unless $dA_0/dx$ is zero. It follows that $A_0$ is a constant.

By (47), $1/y$ is the derivative of $A_1 y$, which is a function uniform on the surface of $y$. Then $A_1 y$ cannot have a pole at a finite point $x_0$. If it did, $1/y$ would have an expansion for $x = x_0$ beginning with a term $a_0(x - x_0)^{-p}$ with $p > 1$. The only places where $1/y$ can show negative exponents are the branch points, where the exponents are $-1/2$. If $A_1 y$ had a pole at $\infty$, $1/y$ would have an expansion at $\infty$ beginning with a nonnegative power of $x$. Thus (47) cannot hold and $I_1$ is not elementary.

We examine $I_2$. The integrand, $x^2/y$, has zero residues at $\pm 1, \pm k$. To find the expansions of $x^2/y$ at $\infty$, we write

(49) $$\frac{x^2}{y} = [k^2 - \frac{1+k^2}{x^2} + \frac{1}{x^4}]^{-\frac{1}{2}} = \pm(\frac{1}{k} + \frac{1+k^2}{2k^3 x^2} + \ldots).$$

Thus the residues at $\infty$ are 0. We have thus to show that $I_2$ is not algebraic. Writing $I_2 = A_0 + A_1 y$, we prove as above that $A_0$ is a constant. Then $x^2/y$ is the derivative of $A_1 y$. As above, $A_1 y$ can have no pole at a finite place. This means that $A_1$ has no poles in the finite complex plane; if it did, such poles could not be removed by multiplication by $y$. Thus $A_1$ is a polynomial. If $A_1$ is of degree $m$, the expansions of $A_1 y$ at $\infty$ begin with terms in $x^{m+2}$; the

---

* Note that this does not happen when $k = 0$. That is what allows the integral of $(1 - x^2)^{-\frac{1}{2}}$ to be elementary.

expansions of the derivative of $A_1 y$ begin with $x^{m+1}$. As $m + 1 > 0$, we see from (49) that the derivative of $A_1 y$ cannot be $x^2/y$.

The subject of elliptic integrals of the third kind is a more delicate one. Such an integral is sometimes a logarithm of a rational combination of x and y. Abel, and later Chebyshev and Zolotareff wrote noteworthy papers on this question.*

### CHEBYSHEV'S INTEGRAL

**14.** Chebyshev considered the integral

$$(50) \qquad u = \int x^p (1 - x)^q \, dx$$

where each of p and q is rational and not zero. He proved that, for the integral to be elementary, it is necessary and sufficient that at least one of p, q and p + q be an integer.

We treat first the question of sufficiency. Let $p = r/t$ and $q = s/t$ where r, s, t are relatively prime integers with t positive.

If q is an integer, we put $x = v^t$, and u becomes the integral of a rational function of v. The integral of a rational function can always be obtained in finite terms by the method of partial fractions. When p is an integer, we replace $1 - x$ by $v^t$. When $p + q$ is integral, we write

$$u = \int x^{p+q} \left(\frac{1-x}{x}\right)^q dx$$

and replace $(1 - x)/x$ by $v^t$, again securing a rational integrand.

To treat the sufficiency question, we start by showing that the integrand in (50), which we denote by y, is a function of t branches. We have

$$(51) \qquad y^t - x^r (1 - x)^s = 0.$$

We consider a branch $y_1$ of y analytic in a small simply connected area $\mathfrak{A}$ which contains neither 0 nor 1. Let a be a point in $\mathfrak{A}$. Let $C_1$ be a simple closed curve, passing through a, with 0 interior to $C_1$ and 1 exterior to it. let $C_2$ be a simple closed curve, passing through a, with 1 interior to $C_2$ and 0 exterior to it. If we move around $C_1$, $y_1$ is replaced, according to (51), by $e^{2\pi i r/t} y_1$. A passage around $C_2$ gives $e^{2\pi i s/t} y_1$. As r, s, t are relatively prime, we can determine integers g, h, k such that

$$(52) \qquad gr + hs + kt = 1.$$

* Appel et Goursat, *Fonctions algébriques* (Paris, 1929), I, 354.

Let C be a curve consisting of g turns around $C_1$ and h turns around $C_2$. Then C replaces $y_1$ by

$$e^{2\pi i(gr+hs)/t}\, y_1 = e^{2\pi i(1-kt)/t}\, y_1 = e^{2\pi i/t}\, y_1.$$

Thus t-1 successive turns around C give t-1 new branches of y in addition to $y_1$. By (51) there are no more than t branches. Thus y has precisely t branches.

Suppose now that none of p, q, p+q is an integer.

We shall prove that the residues of y are all zero. Residues can exist only for x = 0, 1, $\infty$. For x = 0, the expansions of $(1-x)^{s/t}$ are all Taylor expansions. Because r/t is fractional, all expansions of y will contain only fractional powers. Thus, when p < 0, the residues of y at 0 are 0. Similarly, at x = 1, there can be only zero residues. For x = $\infty$, we write

$$y = x^{(r+s)/t}\left(-1 + \frac{1}{x}\right)^{s/t}$$

Again there are only fractional powers and the residues are 0.

We have now to show that u is not algebraic. Let

(53) $$u = A_0 + A_1 y + \ldots + A_{t-1} y^{t-1}.$$

We have

(54) $$\frac{d}{dx} A_i y^i = y^i \left(\frac{dA_i}{dx} + \frac{iA_i}{y}\frac{dy}{dx}\right).$$

Now

(55) $$\frac{1}{y}\frac{dy}{dx} = \frac{1}{ty^t}\frac{dy^t}{dx}$$

and the second member of (55) is a rational function of x. Thus, differentiating (53) we have

(56) $$y = B_0 + B_1 y + \ldots + B_{t-1} y^{t-1}$$

where, for each i, $B_i$ is given by the expression in parentheses in (54) and is a rational function of x. Equation (56) must be an identity, else y would satisfy an equation of degree less than t. Thus $y = B_1 y$ and, adding a constant to u if necessary, we may write

(57) $$u = A_1 y.$$

Suppose now that $A_1 = P/Q$ with P and Q relatively prime polynomials. We shall show first that Q is a constant. Suppose that this is not so. Then Q has a zero at some finite point $\underline{a}$. If $\underline{a}$ is neither

0 nor 1, y is not zero at a, so that the second member of (57) will have poles on the surface of y for x = a. Then the derivative of $A_1 y$ which is y, must have poles at a. This cannot be, so that a = 0 or a = 1. Let a = 0. The expansions of u at 0 begin with fractional exponents less than r/t, and we cannot get a first exponent r/t for y when we differentiate u. Thus a ≠ 0. Similarly, we cannot have a = 1, so that Q is a constant.

Then $A_1$ is a polynomial. If $A_1$ is of degree m, the expansions of u at ∞ start with terms of exponent $m + (r + s) t^{-1}$, which is fractional. To get y from u by differentiation, we must have m = 1. Let then

(58)  $$u = (a + b x) y.$$

Unless a = 0, the expansions of the second member of (58) at x = 0 will start with the exponent r/t and we get too low a first power for y when we differentiate (57). Similarly, a + bx must vanish for x = 1. These demands are too great. The necessity is proved.

## Chapter III

## INTEGRATION OF TRANSCENDENTAL FUNCTIONS

### LIOUVILLE'S GENERAL THEOREM

1. When we say that a function $u(x)$ is algebraic *in* functions $y_1(x), \ldots, y_p(x)$, we shall mean that the functions mentioned are analytic at some point and satisfy an equation

$$\alpha_0 u^m + \ldots + \alpha_m = 0$$

where the $\alpha$ are polynomials in the $y$ with constant coefficients.

Liouville [5] gave the following generalization of his theorem of Chapter II. Let $y_1(x), \ldots, y_p(x)$ be functions satisfying differential equations

(1) $$\frac{dy_i}{dx} = f_i(y_1, \ldots, y_p), \quad i = 1, \ldots, p$$

where the $f$ are algebraic functions of the $y$. Let $u(x)$ be algebraic in the $y(x)$ and let the integral of $u$ be elementary in the $y(x)$, that is, obtainable from the $y$ in a finite number of steps by performing algebraic operations and taking exponentials and logarithms. Then

(2) $$\int u\, dx = v_0 + c_1 \log v_1 + \ldots + c_r \log v_r$$

where the $v$ are algebraic in the $y$. If the $f$ are rational, the $v$ can be taken as rational in the $y$. It follows, for instance, that if $w$ is an elementary function and if the integral of $w$ is elementary, the integral has an expression like the second member of (2) with $v$ which are algebraic in the monomials appearing in $w$.

A neat treatment of these questions is furnished by Ostrowski's recent generalization [17] of Liouville's broader theorem to functions which are algebraic over differential fields. We take up this question.

### DIFFERENTIAL FIELDS

2. Let $\mathfrak{U}$ be an area in the plane of the complex variable $x$. Let $\mathfrak{F}$ be a set of functions each of which is meromorphic in $\mathfrak{U}$ and at least one of which is not identically zero. We shall call $\mathfrak{F}$ a *differential field* if $\mathfrak{F}$ is closed with respect to rational operations and to differentiation. If $f$ and $g$ are functions in $\mathfrak{F}$, $f + g$, $f - g$

and f g are in ℑ. If g is not identically zero, f/g is in ℑ. Again, if f is any function in ℑ, the derivative of f is in ℑ. Thus is made explicit the meaning of the two types of closure.

For instance, the set of all rational functions of x is a differential field. Here 𝔘 is the entire plane.*

### FUNCTIONS ELEMENTARY OVER A FIELD

3. In what follows, we deal with a differential field ℑ which contains all complex constants.

A function u will be said to be *algebraic over* ℑ, or of *order zero over* ℑ, if it satisfies an equation

(3) $$\alpha_0 u^p + \ldots + \alpha_p = 0$$

with the $\alpha$ in ℑ and $\alpha_0$ not zero. To be precise, (3) is satisfied by some element $P(x - x_0)$ of u, and u is taken as the totality of elements obtained by continuing P from $x_0$ along curves lying in 𝔘. Thus, if 𝔘 is not the entire plane, u may be not a complete m.a.f. but only part of such a configuration. This, of course, is an annoyance, but it is not fatal to our theory. In specific applications, our confinement to an area is not a serious matter.

By a *complete monomial of order* 1, *over* ℑ, we shall mean a function which is not of order 0 over ℑ, and which is the exponential or a logarithm of a function of order 0 over ℑ. Again we use what may be a portion of an m.a.f. A *monomial* of order 1 is a branch of a complete monomial analytic in some area in 𝔘. The definitions continue as in Chapter I. We obtain the functions *elementary over* ℑ. They have the property of fluency in the area 𝔘.

A function of order n over ℑ is described by a scheme like that of I, 11, with a slight modification. In (I), we use $r_1$ functions of order 0 over ℑ. In (II), we use $r_2$ functions of x and $x_1', \ldots, x_{r_1}'$, algebraic in the $x_i'$ over ℑ. Such a function satisfies an equation whose coefficients are polynomials in the x' with coefficients in ℑ. Similar remarks apply to (III), ..., (N + 1).

The differentiation of a function elementary over ℑ is discussed as in I, 15. Because ℑ is closed with respect to differentiation, the derivative of a function of x which is algebraic over ℑ is

---

* No generality would be gained by allowing the functions in ℑ to have isolated essential singularities as well as poles. If f(x) has an isolated essential singularity at a, there is, by Picard's theorem, a rational value c which is assumed by f(x) in every neighborhood of a. Then 1/(f(x)-c) has a pole in every neighborhood of a.

algebraic over $\mathfrak{I}$. If a function f of several $x_j^{(1)}$ and x is algebraic in the $x_j^{(1)}$ over $\mathfrak{I}$, each of its partial derivatives is algebraic in the $x_j^{(1)}$ over $\mathfrak{I}$; this is seen with the help of the algebraic equation satisfied by f. Thus the theorem of I, 15, on the differentiation of a function of order n carries over to the present case. One uses a function which is algebraic in the accented letters over $\mathfrak{I}$.

## OSTROWSKI'S GENERALIZATION

4. We may now state the Liouville-Ostrowski theorem, which uses a field $\mathfrak{I}$ containing all complex constants.

THEOREM: *Let* $y(x)$ *be algebraic over* $\mathfrak{I}$. *Let the integral of* $y(x)$ *be elementary over* $\mathfrak{I}$. *Then*

(4) $$\int y\,dx = v_0 + c_1 \log v_1 + \ldots + c_r \log v_r$$

*where the* v *are algebraic over* $\mathfrak{I}$. *Furthermore, the* v *can be taken so as to be rational in* y *and in functions of* $\mathfrak{I}$.

Let u, the integral of y, be of order n > 0 over $\mathfrak{I}$, and let it be expressed with a least number of n-monomials. Let the function, algebraic over $\mathfrak{I}$, associated with u, be

$$g(x_1^{(n)}, \ldots, x)$$

Let each accented variable except $x_1^{(n)}$ be replaced by its monomial. There results a function $f(x_1^{(n)}, x)$. Let $\theta$ represent $\theta_1^{(n)}$. We have

$$u = f(\theta, x).$$

Then

(5) $$y = f_\theta(\theta, x)\,\theta' + f_x(\theta, x).$$

The partial derivatives in (5) are obtained by replacements from functions algebraic in the accented letters over $\mathfrak{I}$. We now follow with no essential change §§ 3, 4, 5 of Chapter II. We secure the relation (4). In § 6 we treat the rationality question.

## OSTROWSKI'S METHOD OF FIELD EXTENSIONS

5. We have followed above the Liouville proof pattern. Ostrowski uses a method which contains a genuinely novel idea and constitutes a noteworthy addition to Liouville's technique. We shall give an account of it.

We consider $\mathfrak{I}$ as above. Let there be given any finite set of functions $u_1, \ldots, u_p$ which are meromorphic in $\mathfrak{U}$ and algebraic over

$\Im$. Let $\Im'$ be the totality of rational combinations of the u and functions in $\Im$. Then $\Im'$ is a differential field, because the derivative of each $u_i$ is rational in that $u_i$ and functions in $\Im$. We call $\Im'$ an *algebraic extension* of $\Im$. By I, 12, every function in $\Im'$ is algebraic over $\Im$.

Now let u be analytic in $\mathfrak{A}$ and algebraic over $\Im$. Suppose that $e^u$ is not algebraic over $\Im$. Let $\Im'$ be the differential field consisting of all rational combinations of $e^u$ and u' with coefficients in $\Im$.* We call $\Im'$, or any algebraic extension of $\Im'$, an *exponential extension* of $\Im$. Similarly, let u, algebraic over $\Im$, be analytic in $\mathfrak{A}$ and suppose besides that log u is uniform in $\mathfrak{A}$ but not algebraic over $\Im$. Let $\Im'$ be the differential field consisting of all rational combinations of log u and u'/u with coefficients in $\Im$. We call $\Im'$, or any algebraic extension of $\Im'$, a *logarithmic extension* of $\Im$.

Given two distinct differential fields $\Im$ and $\Im_1$, if there exists a finite sequence of differential fields,

$$\Im, \Im', \ldots, \Im^{(n)} = \Im_1,$$

each after the first an exponential or logarithmic extension of its predecessor, we call $\Im_1$ an *extension of $\Im$ of positive rank*. If $\Im_1$ is such an extension of $\Im$ and if n, as above, is the smallest number of extensions which will permit the passage from $\Im$ to $\Im_1$, we call n the *rank of $\Im_1$ with respect to $\Im$*.

Now let y(x) be algebraic over $\Im$ and let the integral u of y be elementary, but not algebraic, over $\Im$. From our study of the structure of u, it is seen that we can find an area $\mathfrak{A}_1$ and an extension $\Im_1$ of $\Im$, relative to $\mathfrak{A}_1$ which contains u. After forming an algebraic extension of $\Im$, one brings in, in succession, monomials analytic in $\mathfrak{A}_1$ and functions algebraic in such monomials. Having such an area $\mathfrak{A}_1$ in which extensions containing u exist, we use an extension $\Im_1$ whose rank, r, is as small as possible. For all we can say offhand, r may depend on the area $\mathfrak{A}_1$ which is chosen; r will not increase if $\mathfrak{A}_1$ is shrunk. We suppose $\mathfrak{A}_1$ to be such that r is as small as possible. Let the successive extensions which produce $\Im_1$ from $\Im$ be $\Im'$, $\Im''$, ..., $\Im^{(r)} = \Im_1$. We shall show that the extensions are all *logarithmic* extensions and that

$$u = v_0 + c_1 \log v_1 + \ldots + c_r \log v_r$$

with each $v_i$ algebraic over $\Im$.

---

\* Note that as u' is algebraic over $\Im$, the higher derivatives of u are rational in u' and functions in $\Im$.

First, let $r = 1$. Denoting by $\theta$ the exponential or logarithm used in building $\mathfrak{J}'$ out of $\mathfrak{J}$, we have

$$u = v(\theta, x)$$

with $v$ algebraic in $\theta$ and functions in $\mathfrak{J}$. By the Liouville procedure, we find that

$$u = v_0 + c_1 \log v_1$$

with $v_0$ and $v_1$ algebraic over $\mathfrak{J}$. Suppose now that our theorem is proved for $r = 1, \ldots, s$. Let $r = s + 1$. We simply replace $\mathfrak{J}$ by $\mathfrak{J}'$. $\mathfrak{J}^{(r)}$ is of rank $s$ over $\mathfrak{J}'$. Then

$$u = v_0 + c_1 \log v_1 + \ldots + c_s \log v_s$$

with each $v$ algebraic over $\mathfrak{J}'$ and hence algebraic in the $\theta$ used above and functions of $\mathfrak{J}$. The procedure of Liouville then gives our theorem.

Ostrowski's method of field extensions lends an intriguing Galoisian aspect to the theory of Liouville. As there are no groups in the Liouville theory, the resemblance is not too deep. Ostrowski's method consists, broadly speaking, in expressing $u$ with a minimum total number of monomials of all orders.

## GENERALIZATION OF ABEL'S THEOREM

6. We shall now show that the $v$ in (4) may be taken rational in $y$ and functions in $\mathfrak{J}$. Our method will be algebraic, rather than analytic as in II, 8.

First we secure a function $t(x)$, linear in the $v$ with constant coefficients such that the $v$ are rational in $t$ and functions in $\mathfrak{J}$.

Rather than the $r + 1$ functions $v$, we consider two functions $\varphi$ and $\psi$, analytic at a point $\underline{a}$ in $\mathfrak{U}$ and algebraic over $\mathfrak{J}$. One will see that our conclusion holds for any number of functions. Let $\varphi$ and $\psi$ satisfy equations with coefficients in $\mathfrak{J}$,

$$\alpha_0 \varphi^p + \ldots + \alpha_p = 0; \quad \beta_0 \psi^q + \ldots + \beta_q = 0,$$

the equations being irreducible in $\mathfrak{J}$. These equations have solutions

$$\varphi_1, \ldots, \varphi_p; \quad \psi_1, \ldots, \psi_q$$

analytic in some area close to $\underline{a}$. The $\varphi_i$ are distinct; so also are the $\psi_i$. We understand that $\varphi_1 = \varphi$, $\psi_1 = \psi$.

We wish to choose constants h and k in such a way that the pq functions $h\varphi_i + k\psi_j$ are distinct. The difference of two such functions is of the form $h\zeta + k\xi$ with $\zeta$ and $\xi$ not both zero. The product of all differences of pairs is a polynomial in h and k. If h and k are taken as constants for which the product is not zero identically in x, the pq functions $h\varphi_i + k\psi_j$ will be distinct. Now let

$$F(z) = \prod[z - (h\varphi_i + k\psi_j)], \quad i = 1, \ldots, p; \quad j = 1, \ldots, q.$$

The coefficients in F are symmetric in the $\varphi$ and $\psi$ and are thus in $\mathfrak{J}$. For each i and j, we have

$$F(z) = [z - (h\varphi_i + k\psi_j)] G_{ij}(z)$$

with $G_{ij}$ a polynomial in z. Let

$$H(z) = \Sigma G_{ij}(z) \varphi_i, \quad i = 1, \ldots, p; \quad j = 1, \ldots, q.$$

The coefficients of $H(z)$, by their symmetry, belong to $\mathfrak{J}$. Now, if $F'(z)$ is the derivative of F with respect to z, we have

$$G_{11}(h\varphi_1 + k\psi_1) = F'(h\varphi_1 + k\psi_1),$$

an equation in which the members are not zero, since $F(z)$ has pq distinct factors. For i and j not both unity,

$$G_{ij}(h\varphi_1 + k\psi_1) = 0.$$

Thus

$$H(h\varphi_1 + k\psi_1) = F'(h\varphi_1 + k\psi_1) \varphi_1.$$

Then, if $t = h\varphi + k\psi$, $\varphi$ is rational in t. So also is $\psi$.

7. We suppose that we have a t as indicated for the v. By I, 12, t is algebraic over $\mathfrak{J}$. Let $\mathfrak{D}$ be the totality of rational combinations of y and functions in $\mathfrak{J}$. $\mathfrak{D}$ is a field in the sense of algebra. As $\mathfrak{D}$ contains $\mathfrak{J}$, t satisfies equations with coefficients in $\mathfrak{D}$. Let t satisfy

(6) $$\alpha_0 t^p + \ldots + \alpha_p = 0$$

with coefficients in $\mathfrak{D}$ and irreducible in $\mathfrak{D}$. Equation (6) will have p distinct solutions $t_1, \ldots, t_p$, analytic in some area close to the point $\underline{a}$ at which we are given (4). We let $t_1 = t$. From (2) we have

(7) $$y = v_0' + c_1 \frac{v_1'}{v_1} + \ldots + c_r \frac{v_r'}{v_r}.$$

Now let $v_i = R_i(t, x)$, each R rational in t with coefficients in $\mathfrak{J}$. Then

$$v_i' = \frac{\partial R_i(t, x)}{\partial t} t' + \frac{\partial R_i(t, x)}{\partial x}.$$

By (6) t' is rational in t and functions in $\mathfrak{D}$. On this basis, the second member of (7) may be regarded as a rational combination of t and functions in $\mathfrak{D}$. As (6) is irreducible in $\mathfrak{D}$, we may replace t in (7) by $t_2$, ..., $t_p$ in succession and (7) will continue to hold. This is because an equation with coefficients in $\mathfrak{D}$ which admits one solution of (6) admits all p solutions. We have thus for $j = 1, \ldots, p$,

(8)  $y = \frac{d}{dx} [R_0(t_j, x) + c_1 \log R_1(t_j, x) + \ldots + c_p \log R_p(t_j, x)]$.

We add the p equations (8), observing that $\Sigma R_0(t_j, x)$ is rational in y and functions in $\mathfrak{J}$, as is also the product, for any i, of the $R_i(t_j, x)$, $j = 1, \ldots, p$. We secure a relation (4) with $v_i$ which are rational in y and functions of $\mathfrak{J}$.

A comparison of the above proof with that for the algebraic case in II, 8, brings out an important principle. Where one uses, in an analytical treatment of a problem, the process of analytic continuation, one employs, in an algebraic treatment, the theorem which states that, if $f = 0$ is an irreducible equation and if $g = 0$ admits one solution of $f = 0$, then $g = 0$ admits every solution of $f = 0$. The algebraic theorem just stated is a counterpart of the analytical principle of the permanence of functional equations.

### INTEGRALS OF ELEMENTARY FUNCTIONS

8. Let y in (4) be an elementary function as in Chapter I. Let $\mathfrak{J}$ be the differential field generated by x and y, that is, the totality of rational combinations of x, y, and derivatives of y, with constant coefficients. If the integral u of y is elementary, u will be elementary over $\mathfrak{J}$. It will be expressed as in (4) with each v rational in x, y and a certain number of derivatives of y. Now y and its derivatives are algebraic in the monomials in y. Thus, *if the integral of an elementary function y is elementary, the integral can be expressed as in (4), with each v algebraic in the monomials in y.*

## APPLICATIONS

9. We now consider special types of functions. Liouville investigated the integral

(9) $$u = \int e^{g(x)} y(x)\, dx$$

where g and y are algebraic, with g nonconstant. One works in an area in which definite branches of g and y are taken. Liouville proved that u, when elementary, has the form

(10) $$e^{g(x)} w(x) + c$$

with w rational in x, g, and y.

To secure this result, we let $\mathfrak{I}$ be the totality of rational combinations of x, $e^{g(x)}$, g, and y. Then $\mathfrak{I}$ is a differential field, since $g'$, for instance, is rational in x and g. Let u be elementary. Let $\theta = e^{g(x)}$. Then

(11) $$u = v_0(\theta, x) + \Sigma\, c_i \log v_i(\theta, x)$$

with each v rational in $\theta$, x, g, y. Differentiating, we have

(12) $$\theta y = \frac{\partial v_0}{\partial \theta}\, \theta g' + \frac{\partial v_0}{\partial x} + \Sigma\, \frac{1}{v_i}\left[\frac{\partial v_i}{\partial \theta}\, \theta g' + \frac{\partial v_i}{\partial x}\right]$$

As $\theta$, by I, 5, is transcendental, (12) is an identity in $\theta$. We replace $\theta$ by $\mu\theta$ and integrate. Then

$$\mu u = v_0(\mu\theta, x) + \Sigma \log v_i(\mu\theta, x) + \beta(\mu).$$

We differentiate with respect to $\mu$ and put $\mu = 1$. We find for u an expression rational in $\theta$, x, g, y. Let such an expression be given by

$$u = v(\theta, x).$$

We find as above

$$v(\mu\theta, x) = \mu\, v(\theta, x) + \beta(\mu).$$

Differentiation with respect to $\mu$ gives, for $\mu = 1$,

$$\theta\, \frac{\partial v(\theta, x)}{\partial \theta} = v(\theta, x) + c.$$

We replace $\theta$ by a variable z and integrate the resulting equation. We find that

$$v(z, x) = -c + z\,\gamma(x) = -c + \frac{z}{z_0}[v(z_0, x) + c].$$

Thus

$$u = \theta\left[\frac{v(z_0, x) + c}{z_0}\right] - c$$

so that u is expressed as in (10).

10. It is noteworthy that the manner of pairing the branches of g and y may affect the character of the integral. Thus if

$$g = (1 + x^{\frac{1}{2}})^2, \quad y = 1 + x^{-\frac{1}{2}}$$

where the same branch of $x^{\frac{1}{2}}$ is used in g and y, the integral is simply $e^g$. If we use one branch in g and the other, its negative, in y, then u is not elementary. If it were, we would find by subtraction that the integral of $x^{-\frac{1}{2}} e^g$ is elementary. Putting $z = 1 + x^{\frac{1}{2}}$, we would find the integral of $e^{z^2}$ to be elementary. This, it will be seen below, is not so.

11. To develop a practical method for determining whether (9) is or is not elementary, it suffices to consider the case of $g(x) = x$. For u to be elementary, the differential equation

(13) $$w' + w = y$$

must have a solution rational in x and y. The treatment of this question parallels the work of II, 10. We shall examine the case in which y is rational. In that case, if (13) has a rational solution w, the poles of w in the finite plane must be poles of y. If y has a pole of order p at a point $\underline{a}$, w must have at $\underline{a}$ a pole of order p - 1. Thus if y = P/Q with P and Q polynomials, and if w is a rational solution of (13), Qw will be a polynomial. We put w = z/Q and (13) goes over into

(14) $$Q z' + (Q - Q')\,z = PQ.$$

We have to see whether (14) has a solution z which is a polynomial. The degree of Q - Q' is that of Q. It follows that z must have the same degree as P. Thus, if P is of degree m, we substitute a general polynomial of degree m for z in (14) and secure a system of linear equations whose compatibility or incompatibility decides the character of our integral.

12. Let, for example,

(15) $$u = \int e^{-x^2}\,dx.$$

If u were elementary, it would be given by (10) where w is rational and satisfies

(16) $\qquad w' - 2xw = 1.$

As the second member of (16) has no pole, no solution of (16) can have a pole in the finite part of the plane. Furthermore, w cannot be a polynomial. If it were, xw would be of higher degree than w' and 1 and (16) could not hold. Thus the integral (15) is not elementary.

Consider now the integral

(17) $\qquad u = \int \dfrac{e^x}{x} dx.$

For it to be elementary, there would have to be a rational w with

$$w' - \dfrac{w}{x^2} = \dfrac{1}{x}.$$

Let such a rational solution exist. Let its expansion at the origin be

$$w = a_0 x^p + a_1 x^{p+1} + \ldots.$$

We find that $p = 1$. Thus w is a polynomial of positive degree. For the neighborhood of $\infty$, the expansion of w' begins with a higher power than does that of the other terms. We see that (17) is not elementary.

We consider finally the integral

$$u = \int \dfrac{dx}{\log x}$$

which occurs in the theory of the distribution of prime numbers. Putting $x = e^t$ we find the integral to go over into (17). It is thus not elementary.

13. Liouville extended, as follows, the result presented in §9. Let

(18) $\qquad w = e^{g_1} y_1 + \ldots + e^{g_p} y_p,$

*with algebraic g and y, have an elementary integral. Suppose further that no two $g_i$ differ by a constant. Then each $e^{g_i} y_i$ has an elementary integral.* It is understood that one of the g may be a constant.

We assume, as we may, that $p > 1$. Our first step will be to secure functions $h_1, \ldots, h_q$ with $1 \leq q \leq p$ which satisfy the following conditions:

(a) There exists no relation $m_1 h_1 + \ldots + m_q h_q + c = 0$, with c a constant and the m integers not all zero.

(b) Each g differs by a constant from a linear combination, with integral coefficients, of the h.

(c) Each h is some g divided by an integer.

Let $p = 2$. If there exists no relation $m_1 g_1 + m_2 g_2 + c = 0$ with $m_1$ and $m_2$ integers, we take $q = 2$, $h_1 = g_1$, $h_2 = g_2$. Let such a relation exist and, to fix our ideas, suppose that $m_2 \ne 0$. We notice that $g_1$ is not a constant. If it were, $g_2$ would be a constant and so also would be $g_2 - g_1$. We take $q = 1$ and $h_1 = g_1/m_2$.

Suppose then that we have treated all values of p with $p \le r$. We examine the case of $p = r + 1$. For $g_1, \ldots, g_r$, we have a set of functions satisfying (a), (b), (c). We denote them by $h_1', \ldots, h_{q'}'$. If $g_{r+1}$ does not differ by a constant from a linear combination of the other g with rational coefficients, we take $q = q' + 1$, $h_i = h_i'$, $i = 1, \ldots, q'$, and $h_q = g_{r+1}$. Now, suppose that

$$g_{r+1} = \frac{m_1}{n} g_1 + \ldots + \frac{m_r}{n} g_r + c.$$

We take $q = q'$, $h_i = h_i'/n$, $i = 1, \ldots, q'$.

The property (c) will not be used in what follows; it was important only for guaranteeing (a) at each step of the induction.

Now let $h_1, \ldots, h_q$ be any finite set of algebraic functions satisfying (a) above and not necessarily derived, as in what precedes, from a system $g_1, \ldots, g_p$. Let $\theta_i = e^{h_i}$, $i = 1, \ldots, q$. We shall prove that there exists no algebraic relation

(19)   $$F(x, \theta_1, \ldots, \theta_q) = 0$$

which is not an identity in the x and the $\theta$. This is true for $q = 1$, since $\theta_1$ is not algebraic. We examine the case of $q = r + 1$, supposing the lower cases to have been treated. Suppose that (19) with $q = r + 1$ holds but is not an identity. Then $\theta_q$ must figure effectively in (19), else we would have an algebraic relation among $x, \theta_1, \ldots, \theta_r$. Thus

(20)   $$\theta_{r+1} = G(x, \theta_1, \ldots, \theta_r)$$

with G algebraic.* We take logarithmic derivatives in (20). Then

(21)   $$h_{r+1}' = \frac{1}{G} [G_x + \frac{\partial G}{\partial \theta_1} \theta_1 h_1' + \ldots + \frac{\partial G}{\partial \theta_r} \theta_r h_r'].$$

* The details are as in I, 14.

In (21), we have an identity in x, $\theta_1$, ..., $\theta_r$. By the μ-process, we find

$$\theta_{r+1} = e^{\beta(\mu)} G(x, \mu\theta_1, \theta_2, ..., \theta_r)$$

with $\beta(\mu)$ analytic for $\mu = 1$. Then, for a variable z and a constant c,

$$z \frac{\partial G(x, z, \theta_2, ..., \theta_r)}{\partial z} + c\, G(x, z, \theta_2, ..., \theta_r) = 0.$$

Then

(22) $G(x, z, \theta_2, ..., \theta_r) = z^{-c} \gamma(x) = z^{-c} z_0^c\, G(x, z_0, \theta_2, ..., \theta_r).$

As G is algebraic in z, c is rational. By (20) and (22)

(23) $\qquad \theta_{r+1}\, \theta_1^c = e^{h_{r+1} + c\, h_1} = G(x, z_0, \theta_2, ..., \theta_r).$

Now (23) is a nonidentical algebraic relation among the exponentials of the r functions $h_{r+1} + c h_1, h_2, ..., h_r$, which latter satisfy (a) above. Our statement is proved.

We return now to $g_1, ..., g_r$ as given above and to $h_1, ..., h_q$ as found for them, satisfying (a) and (b). We subtract constants from the g in such a way that each of them becomes a linear combination of the h with integral coefficients. This is balanced by multiplying the y in (18) by constants. Supposing the integral u of w in (18) to be elementary, we have

$$u = v_0 + \Sigma\, c_i \log v_i$$

with each v algebraic in x, $\theta_1, ..., \theta_q$ where $\theta_i = e^{h_i}$.

Now w in (18) is a sum of terms each of which is a product of an algebraic function and integral powers of the θ. The customary procedure of Liouville shows that if we replace any θ in w by a constant times itself, we get a function which is integrable in finite terms. Two terms of w involve the θ with distinct sets of exponents. Let us separate w into sets of terms, all terms of any one set involving $\theta_1$ to the same power and distinct sets showing $\theta_1$ in different powers. Let w, thus written, be

(24) $\qquad\qquad w = \theta_1^{m_1} A_1 + ... + \theta_1^{m_s} A_s.$

For every μ, the integral of

$$\mu^{m_1} \theta_1^{m_1} A_1 + ... + \mu^{m_s} \theta_1^{m_s} A_s$$

is elementary. As the m are distinct, we can find numbers $\mu_1, ..., \mu_s$

for which the determinant whose $j^{th}$ now is $\mu_j^{m_1}, \ldots, \mu_j^{m_s}$ does not vanish; this follows from the linear independence, as functions of $\mu$, of the $\mu^{m_1}$. Thus each term in (24) has an elementary integral. We arrange such a term according to powers of $\theta_2$ and continue. We find each term in (18) to be integrable in finite terms.

14. How far does Liouville's theory go? To what extent can one determine whether a given function can be integrated in finite terms? Mordukhai-Boltovskoi has been the chief investigator of this question. In his book [13] many types of functions are examined and the general forms of their integrals, when elementary, are obtained. Numerous special examples are treated in detail. What is of greatest theoretical interest in Mordukhai-Boltovskoi's work is his study of elementary functions in which the algebraic operations employed are all rational. One uses thus $e^x$ and log x, but no irrational algebraic operations. A method is developed for testing for the integrability of such functions. Details are presented for functions of orders 1 and 2.

Mordukhai-Boltovskoi uses three types of monomials. They are exponentials, irrational powers, and logarithms. Thus $x^a$ with $a$ irrational is of the first order in his theory, whereas, as will be seen in IV, 2, it is of the second order in the classification of Liouville. Liouville's procedure has the merit of employing a minimum number of monomials. For instance, as can be shown by the methods, of the following chapter, log x cannot be obtained by performing algebraic operations and taking exponentials. However, the irrational powers have a definite formal utility.

## Chapter IV

## FURTHER QUESTIONS ON THE ELEMENTARY FUNCTIONS

### EXISTENCE OF FUNCTIONS OF ALL ORDERS

1. We shall prove, by means of an example, the existence of functions of order n for every positive integer n. Let $e_1(x)$ denote $e^x$. We let $e_2(x)$ represent the exponential of $e_1(x)$ and, in general, $e_n(x)$ the exponential of $e_{n-1}(x)$. Liouville [6] showed that the order of $e_n(x)$ is n. We shall establish this result.

If $y(x)$ is of order n, $e^{y(x)}$ will be of order n-1 if, and only if, y is an n-logarithm. One might inquire as to the circumstances under which $e^y$ is of order n. We shall prove, in this connection, the following lemma.

LEMMA: *If* $e^{y(x)}$, *with* $y(x)$ *of order* n, *is of order* n, *then*

(1) $$y(x) = u(x) + \log v(x)$$

*with* $\log v(x)$ *an n-logarithm and with* $u(x)$ *of order n-1.*

We express $y(x)$ and its exponential, which latter we call $w(x)$, in such a way that the number of n-monomials which appear in at least one of them is as small as possible. Let θ be one of these monomials. We write

$$y(x) = f(\theta, x), \quad w(x) = g(\theta, x),$$

so that

(2) $$f(\theta, x) = \log g(\theta, x).$$

Let $\theta = e^{p(x)}$ with $p(x)$ of order n-1. We find first that

$$f(\mu\theta, x) = \log g(\mu\theta, x) + \beta(\mu),$$

then that, for a variable z,

(3) $$f(z, x) = \log g(z, x) + c \log z + \gamma(x).$$

We are going to prove that $f(z, x)$ is independent of z, that c is a rational number distinct from zero, and that

$$g(z, x) = z^{-c} w_1(x)$$

with $w_1$ algebraic in the monomials other than θ which figure in w. We shall know thus, in particular, that θ does not figure in y.

Suppose that $f(z, x)$ is not independent of $z$. Then there is a value $x_0$ of $x$, close to the point $x = a$ near which we are working, such that $f(z, x_0)$ is a nonconstant function of $z$. By (3),

(4)  $\qquad f(z, x_0) = \log g(z, x_0) + c \log z + \gamma(x_0).$

We let $h(z) = f(z, x_0) - \gamma(x_0)$ and $k(z) = g(z, x_0)$, so that

(5)  $\qquad h(z) = \log k(z) + c \log z.$

The algebraic functions $h(z)$ and $k(z)$ are given to us through branches analytic at $z = \theta(a)$. By III, 6, we can find a $t(z)$, linear in $h$ and $k$, such that $h$ and $k$ are rational in $t$ and $z$ and thus rational on the Riemann surface of $t$. We have thus, everywhere on the surface of $t$,

(6)  $\qquad h'(z) = \dfrac{k'(z)}{k(z)} + \dfrac{c}{z}.$

As $h(z)$ is not constant, there is a place on the surface of $t$ at which it has a pole. Let such a place lie over $z = b$. First let $b$ be finite. Then $h'(z)$ has an expansion, at the pole, in which the lowest exponent of $t - b$ is less than $-1$. As was seen in II, 11, we cannot get such an exponent from the second member of (6). Let $b = \infty$. Then $h'(z)$ has at $\infty$ an expansion in which the highest exponent exceeds $-1$. Again we find a contradiction. Thus, for every $x_0$ close to $a$, $f(z, x_0)$ is a constant. We take any such $x_0$ and consider (5) where $h(z)$ is a constant. Then (6) becomes

(7)  $\qquad \dfrac{k'(z)}{k(z)} + \dfrac{c}{z} = 0.$

If $c$, which is independent of $x_0$, were zero, $g(z, x_0)$ would be free of $z$ for every $x_0$ and $\theta$ would be absent from $w$ as well as from $y$. Then $c \ne 0$. This means, by II, 11, that $k'/k$ has a residue for $z = 0$ which is a multiple of $c$. As a residue of $k'/k$ must be an integer, $c$ is a rational number. By (7), $k(z) = d\, z^{-c}$ with $d$ constant. Then $z^{-c} g(z, x)$ is independent of $z$. It is a function $w_1(x)$, algebraic in the monomials other than $\theta$ which appear in $w$.

We have $w(x) = \theta^{-c} w_1(x)$. We can now prove that there is no other exponential among the n-monomials in $y$ and $w$. Suppose that there is one, $\zeta = e^{q(x)}$. Let $y_1(x) = y(x) + c\, p(x)$, where $\theta = e^p$. We have

$$e^{y_1(x)} = w_1(x).$$

Now $y_1$, like $y$, is free of $\theta$. Thus the n-monomials which appear in

$y_1$ and $w_1$ are those other than $\theta$ which appear in $y$ and $w$. It follows as above that $\zeta$ does not appear in $y_1$ and that $w_1 = \zeta^{-d} w_2$ with $d$ rational and $w_2$ free of $\theta$ and $\zeta$. As

$$w = e^{-cp-dq} w_2,$$

$y$ and $w$ are expressible with fewer n-monomials than were used above.

We have thus $w = e^{-cp(x)} w_1$ with $w_1$ free of n-exponentials. We shall show now that $w$ contains no n-logarithm. Let $\theta$ be such a monomial. We use (2) and find first

$$f(\theta+\mu, x) = \log g(\theta+\mu, x) + \beta(\mu),$$

then

$$f(z, x) = \log g(z, x) + c_1 z + \gamma(x).$$

If $g(z, x)$ were not free of $z$, we would find, fixing $x$ at a suitable $x_0$, a nonconstant algebraic function whose logarithm is algebraic. Altogether,

$$w = e^{-cp(x)} w_1$$

with $w_1$ of order less than n. Then

$$y(x) = -c\, p(x) + \log w_1(x).$$

Necessarily $\log w_1$ is of order n. The lemma is proved.

We can now show that $e_n(x)$ is of order n. We know that $e_1$ is of order 1. Suppose that we have accounted for the cases from 1 to n-1 inclusive. We make an induction. Since $e_{n-1}$ is of order n-1, $e_n(x)$ is of one of the orders n-2, n-1, n. Let it have order n-2 or n-1. By our lemma,

(8) $$e_{n-1} = u + \log v$$

with $v$ of order n-2 and $u$ of order less than n-1. We note that if $e_n$ is of order n-2, we merely take $v = e_n$ and $u = 0$. Differentiating, we find

$$e_{n-1}\, e_{n-2} \cdots e_1$$

to be of order less than n-1. As the reciprocal of $e_{n-2} \cdots e_1$ is at most of order n-2, $e_{n-1}$ is of order less than n-1. Thus $e_n$ is of order n.

In a similar way, one can prove that the $\underline{n}^{th}$ iterate of $\log x$ is of order n.

## IRRATIONAL POWERS

2. We consider the function $x^a$ with $\underline{a}$ irrational. We shall prove, with Liouville, that it is of order 2. Its expression $e^{a \log x}$ shows that it has an infinite number of branches; consequently it is not algebraic. If it were of order 1, we would have, by §1,

$$a \log x = u + \log v$$

with u and v algebraic. Then

$$\frac{a}{x} = u' + \frac{v'}{v}.$$

This is impossible because $a/x$ has irrational residues, $u'$ only zero residues, and $v'/v$ integral residues.

## KEPLER'S EQUATION

3. Liouville considered the equation of Kepler

$$(9) \qquad y - a \sin y = x$$

with $\underline{a}$ a constant distinct from zero, an equation which occurs in the elements of astronomical theory. He proved that the function $y(x)$ defined by the equation is not elementary.

We write

$$(10) \qquad y = \arcsin \frac{y-x}{a}$$

and recall that

$$\arcsin v = -i \log (iv + \sqrt{1-v^2}).$$

Then (10) becomes, for $z = iy$,

$$(11) \qquad z = \log w(x, z)$$

where w, algebraic in x and z, involves z effectively. Equation (11) is certainly not satisfied by a nonconstant algebraic $z(x)$. Let (11) have an elementary solution $z(x)$ of order $n > 0$. As $w(x, z)$ will be of order n, the lemma of §1 shows that $z = u + \theta$ with $\theta$ an n-logarithm and u of order n-1. Then w, as well as z, can be expressed with the single n-monomial $\theta$ and, when so expressed, will involve $\theta$ effectively. But the work of §1 shows that $\theta$ will not be present in w. Thus $z(x)$ is not elementary.

In a similar way, one can treat

$$u(x, y) = \log v(x, y)$$

with u and v algebraic, at least one of them involving y effectively. If one of u and v is free of y, there will be an elementary solution. If u and v both involve y, we can show, as above, that there is no elementary solution of positive order. The existence of algebraic solutions depends on that of constants c for which the equations $u = \log c$, $v = c$ have a solution $y(x)$ in common.

## INVERSES OF ELEMENTARY FUNCTIONS

4. In connection with Kepler's equation, we met a function of y, namely $y - a \sin y$ whose inverse is not elementary. This raises the question as to when an elementary function has an elementary inverse. The author treated this question [19]. It was shown that when $y(x)$, of order $n > 0$, has an elementary inverse, there exist n monomials of the first order, $\theta_1, \ldots, \theta_n$, such that y is an algebraic function of $\varphi_n$ where

$$\varphi_1 = \theta_1(x), \varphi_2 = \theta_2[\varphi_1(x)], \ldots, \varphi_n = \theta_n[\varphi_{n-1}(x)].$$

That a function of this type has an elementary inverse is obvious.

This result is obtained after there is proved a theorem on functions of functions. The theorem states that if $\varphi(x)$ is of order m and if a $\psi(x)$ of order n exists such that the order of $\psi[\varphi(x)]$ is at most $m + n - 2$, then $\varphi(x)$ is an algebraic function of an n-monomial.

In the proof of this theorem of composition, there is a lemma which throws some light on the structure of elementary functions. By a *logarithmic sum of order* n is meant a function of order n of the form

$$c_1 \log v_1(x) + \ldots + c_r \log v_r(x)$$

with constant c and with v of orders not exceeding n-1. Of course, at least one v is of order n-1. The lemma states that if, in the expression of a function y of order n, the number of n-exponentials plus the number of logarithmic sums of order n is a minimum, each n-exponential and each logarithmic sum of order n is algebraic in y, a certain number of derivatives of y, and the monomials of orders less than n which appear in the expression for y.

One may inquire as to what happens when y is expressed with a minimum total number of exponentials and logarithmic sums of all orders from 1 to n. Will each exponential and each sum be algebraic in y, derivatives of y, and x? An example like $e^{a \log x}$ with $\underline{a}$ irrational shows that the answer is negative. Perhaps to secure an extension of the lemma, one should employ a new type of monomial like, for instance, Mordukhai-Boltovskoi's irrational powers of III, 14.

E. R. Lorch has studied elementary transformations, in two variables, which have elementary inverses [10].

## SCOPE OF METHOD

5. The method of Liouville can be applied to categories of functions built with transcendental operations other than the taking of exponentials and logarithms. The operation of integration will be considered in Chapter VI. At this point we should like to consider the possibility of using functions other than $e^x$ and $\log x$ in forming monomials of the first order. We have already mentioned $x^a$ with $\underline{a}$ irrational. This does not give a larger class of functions; rather, it facilitates the study of some problems on elementary functions.

We notice that $\log x$ is an integral of $1/x$. Now suppose that $f(x)$ is a nonelementary function whose derivative is algebraic, for instance, an elliptic integral of the first kind. To present a simple scene, suppose that we build a class of functions using one transcendental operation, that of taking the function f of an expression. Thus $f(v)$ with v of order n-1 will be an n-monomial if it is not of order lower than n. Let

$$(12) \qquad F(\theta, \ldots, x),$$

of order n, be expressed with the usual economy, and let its derivative be of order less than n. Then

$$F(\theta+\mu, \ldots, x) = F(\theta, \ldots, x) + \beta(\mu)$$

and it is seen that θ figures linearly in F.

Again, let g be a function which is not elementary but whose inverse, like that of $e^x$, is the integral of an algebraic function. Then

$$g' = \varphi(g)$$

with φ algebraic. Suppose that we build monomials using the operation g, with perhaps others, and that (12) has a derivative of order less than n. If $\theta = g(v)$, we have

$$(13) \qquad \frac{d}{dx} F(\theta, \ldots, x) = \frac{\partial F}{\partial \theta} \varphi(\theta) v' + \ldots .$$

We replace $\theta$ in the second member of (13) by $g(v + \mu)$ and find that

$$F(\theta, \ldots, x) = F(g(v + \mu), \ldots, x) + \beta(\mu).$$

We differentiate with respect to $\mu$ and put $\mu = 0$. Then

$$\frac{\partial F}{\partial \theta} \varphi(\theta) = c$$

from which we infer the superfluity of $\theta$ in F.

## HYPERTRANSCENDENTAL FUNCTIONS

6. It is seen without difficulty that every elementary function satisfies an algebraic differential equation, that is, an equation

(14) $$P(x, y, y', \ldots, y^{(n)}) = 0$$

where P is a polynomial with constant coefficients. Thus, a hypertranscendental function, one which does not satisfy an equation (14), is not elementary. For instance, the gamma function, proved hypertranscendental by Hoelder, is not elementary.

## PROBLEMS OF THE THEORY OF FUNCTIONS

7. It is impossible to work long with the elementary functions without finding functiontheoretic problems. For instance, one might undertake to determine all elementary functions which are uniform. This may be a complicated question. The author [22] has examined certain expressions of the first order. An exponential sum is a function

$$c_1 e^{a_1 x} + \ldots + c_r e^{a_r x}$$

with constant c and $\underline{a}$. Let $w(x)$ be defined by an equation

$$\alpha_0 w^p + \ldots + \alpha_p = 0$$

with each $\alpha$ an exponential sum. It is shown that if w is uniform, or more generally, if it is uniform in a sector of opening greater than $\pi$, then w is the quotient of two exponential sums. If the quotient of two exponential sums is an integral function, it is an exponential sum. A factorization theory for exponential sums is obtained. These questions are related to the elegant theory of the zeros of exponential sums developed by Tamarkin, C. E. Wilder, Polya, and Schwengeler.*

* See R. L. Langer, *Bulletin of the American Mathematical Society*, XXXVII (1931), 213.

## RELATED QUESTIONS
### ELEMENTARY NUMBERS

8. We should like to present a class of problems in which numbers are involved rather than functions.

The classification of numbers as algebraic or transcendental is well known. A number is algebraic if it satisfies an algebraic equation with integral coefficients, not all zero; otherwise the number is transcendental.

Let us define *elementary number*. An algebraic number will be called a *number of order zero*. The exponential of any algebraic number distinct from zero, or a logarithm distinct from zero of an algebraic number, will be called a *monomial of order unity*. A *number of order unity* is one which is not algebraic and satisfies an algebraic equation whose coefficients are polynomials, with integral coefficients, in monomials of order one. Continuing, we secure the *elementary numbers*.

There arises immediately, of course, the problem of the existence of numbers of all orders. Priority should perhaps be given to problems on the character of the roots of simple transcendental equations. One might ask, for instance, whether the equation

$$e^z = z$$

has an elementary root. These are, of course, problems of greater difficulty than those which we have been studying.

Chapter V

SERIES OF FRACTIONAL POWERS

1. In the next chapter, there will be presented a variation of Liouville's technique for the treatment of problems of integration in finite terms. The new procedure will require us to develop a function $f(x, y)$, algebraic in y, in descending powers of y for the neighborhood of infinity. A brief derivation of developments of this type, based on the Weierstrass preparation theorem, has been given by Ostrowski.* We use here rather a version of the Newton polygon process, gaining perhaps, in the self-contained nature of our treatment, what is lost as regards brevity.

2. We work first in the neighborhood of $y = 0$. We deal with an equation

(1)    $A_n(x, y) z^n + A_{n-1}(x, y) z^{n-1} + \ldots + A_0(x, y) = 0$

with $n \geq 1$, where each A is analytic in x and y for every x in a given area $\mathfrak{U}$ and for every y with $|y| \leq \eta$ where $\eta > 0$. We understand that $A_n$ does not vanish identically in x and y.

We assume that (1) is irreducible in the (algebraic) field of the A.

For our purposes, it is desirable to replace (1) by an equation in which the coefficient of $z^n$ is unity. Accordingly, we expand each A in a series of powers of y and divide by $A_n$. Equation (1) takes the form

(2)    $F(x, y, z) = z^n + B_{n-1}(x, y) z^{n-1} + \ldots + B_0(x, y) = 0$,

where $B_i$, if not identically zero, has an expansion

(3)    $B_i = a_{0i}(x) y^{\sigma_i} + a_{1i}(x) y^{\sigma_i + 1} + \ldots$

with $\sigma_i$ an integer, positive, negative, or zero, and with $a_{0i}(x)$ not zero. All series (3) converge for x in some area $\mathfrak{B}$ contained in $\mathfrak{U}$ and for every $y \neq 0$ with $|y| < \eta_1$, where $\eta_1 \leq \eta$. Equation (2) is irreducible in the field of the B.

We are going to show that $F(x, y, z)$ in (2) has a representation

$$\prod_{i=1}^{n} [z - P_i(x, y)]$$

* *Mathematische Zeitschrift*, XXXVII (1933), 101.

in which each P is a series of the type

(4) $$c_1(x) y^{p_1} + c_2(x) y^{p_2} + \ldots$$

where the p are rational numbers, with a common denominator, which increase with their subscripts. The p and the $c(x)$ will depend upon the subscript i of $P_i$. The n series (4) will converge for every x in some area and for $|y|$ small and distinct from 0.

3. We shall prove first the existence of a single series (4). After that it will be easy to get n such series.

If $B_0 = 0$, the quantity 0 is a series (4) which answers our requirements.* In what follows, we assume that $B_0$ is not identically zero. We consider the ratio

(5) $$\frac{\sigma_0 - \sigma_i}{i}$$

which exists for every $i \geq 1$ for which $B_i \neq 0$, in particular for $i = n$.** Let $p_1$ be the greatest of the ratios (5).

Let $g_1$ be the greatest value of i for which (5) equals $p_1$. For $i = 0, 1, \ldots, g_1$, we define a function $k_i'(x)$ as follows.

We let $k_0'(x) = a_{00}(x)$. For $i > 0$, if $B_i \neq 0$ and if (5) equals $p_1$, we let $k_i'(x) = a_{01}(x)$. If, for an $i > 0$, either $B_i = 0$ or else (5) is less than $p_1$, we let $k_i'(x) = 0$.

We consider the equation for an unknown function $c(x)$,

(6) $$k_0'(x) + k_1'(x) c + \ldots + k_{g_1}'(x) c^{g_1} = 0.$$

An area contained in $\mathfrak{B}$ can be found in which (6) has $g_1$ analytic solutions, not necessarily distinct. Let $c_1(x)$ be any one of these solutions. For later purposes we shall select in a special way an area in which $c_1(x)$ will be studied. Let $c_1(x)$ be a solution of (6) of multiplicity $s_1$. The $\underline{s_1}^{th}$ derivative with respect to c of the first member of (6) does not vanish identically in x for $c = c_1(x)$. We shall work with $c_1(x)$ in an area $\mathfrak{B}_1$ throughout which the above $\underline{s_1}^{th}$ derivative is distinct from zero.

4. It may be that $c_1(x) y^{p_1}$ causes $F(x, y, z)$ in (2) to vanish identically in x and y when substituted for z. If so, $c_1 y^{p_1}$ is a series (4) such as we are seeking. In what follows, we assume that the vanishing does not occur.

We put $z = c_1 y^{p_1} + z_1$ in (2). Then

* This can happen only if $n = 1$.
** We understand that $B_n = 1$.

(7) $$F(x, y, z) = F_1(x, y, z_1)$$

where $F_1$ is a polynomial of degree n in $z_1$. We may write

(8) $$F_1 = B'_n z_1^n + B'_{n-1} z_1^{n-1} + \ldots + B'_0 = 0.$$

Each $B'$ is a series of ascending rational powers of y, the exponents of y having as their denominator the denominator of $p_1$. The coefficients in the $B'$ are analytic in $\mathfrak{B}_1$. $B'_0$ is not identically zero, since $c_1 y^{p_1}$ does not annul F. As to $B'_n$, it is merely unity.

For every i for which $B'_i \neq 0$, we let $\sigma'_i$ denote the least exponent of y in $B'_i$. There is a $\sigma'_0$ and a $\sigma'_n$, the latter being 0. We denote by $p_2$ the greatest of the quantities

(9) $$\frac{\sigma'_0 - \sigma'_i}{i}$$

5. We are going to prove that $p_2 > p_1$. We shall make, in F of (2), the substitution $z = c y^{p_1}$ where c is an indeterminate. F becomes a collection of terms of the type $c^i a(x) y^b$, where b is rational. We shall show that the least b is $\sigma_0$ and we shall find the terms with $b = \sigma_0$. From $B_i z^i$ with $B_i \neq 0$, the lowest term secured is

$$a_{oi}(x) c^i y^{\sigma_i + i p_1}$$

If $i = 0$, or if $i > 0$ and (5) equals $p_1$, $\sigma_i + i p_1$ equals $\sigma_0$. If (5) is less than $p_1$, $\sigma_i + i p_1$ exceeds $\sigma_0$. Thus the least b is $\sigma_0$, and the sum of the terms with $b = \sigma_0$ may be written

$$y^{\sigma_0} [k'_0(x) + k'_1(x) c + \ldots + k'_{g_1}(x) c^{g_1}].$$

The bracket in this expression will be designated by $\varphi_1(x, c)$.

If now $\tau$ is the least value of b which exceeds $\sigma_0$, we may write

(10) $$F(x, y, c y^{p_1}) = \varphi_1(x, c) y^{\sigma_0} + \psi_1(x, y, c) y^\tau$$

where $\psi_1$ is a polynomial in c whose coefficients are series of terms of the form $a(x) y^d$ with d rational and nonnegative.

In (10), we put

$$c = c_1(x) + z_1 y^{-p_1}.$$

The first member of (10) becomes $F(x, y, c_1 y^{p_1} + z_1)$, which equals $F_1(x, y, z_1)$. Hence

(11) $$F_1(x, y, z_1) = \varphi_1(x, c_1 + z_1 y^{-p_1}) y^{\sigma_0} + \psi_1(x, y, c_1 + z_1 y^{-p_1}) y^\tau.$$

Expanding in powers of $z_1$, we find

(12) $$B_i^! = y^{\sigma_0 - i p_1} \frac{\varphi_1^{(i)}(x, c_1)}{i!} + y^{\tau - i p_1} \frac{\psi_1^{(i)}(x, y, c_1)}{i!}$$

where the superscript (i) denotes i differentiations with respect to c.

Let us consider a $B_i^!$ which is not zero. As $\tau > \sigma_0$, we find that if $\varphi_1^{(i)}(x, c_1)$ does not vanish identically in x, we have

(13) $$\sigma_i^! = \sigma_0 - i\, p_1$$

but that, if the vanishing does occur,

(14) $$\sigma_i^! > \sigma_0 - i\, p_1.$$

In particular, (14) holds for i = 0, so that $\sigma_0^! > \sigma_0$.

On the other hand, $s_1$ being, as at the end of §3, the multiplicity of the solution $c_1(x)$ of (6), we find (13) to hold for $i = s_1$.

We have

(15) $$\frac{\sigma_0^! - \sigma_i^!}{i} = \frac{\sigma_0^! - \sigma_0}{i} + \frac{\sigma_0 - \sigma_i^!}{i}.$$

For $i = s_1$, the second term in the second member of (15) is $p_1$. The first term in the second member is positive, since $\sigma_0^! > \sigma_0$. Thus $p_2$, the greatest value of (9), exceeds $p_1$.

6. Let $g_2$ be the greatest value of i for which (9) equals $p_2$. We denote by $k_0''(x)$ the coefficient of $y^{\sigma_0^!}$ in $B_0^!$. For $i > 0$, if $B_i^! \ne 0$ and if (9) equals $p_1$, we let $k_i''(x)$ denote the coefficient of $y^{\sigma_i^!}$ in $B_i^!$. In the other cases with $i > 0$, we let $k_i''(x)$ denote 0. We consider the equation

(16) $$k_0''(x) + k_1''(x)\, c + \ldots + k_{g_2}''(x)\, c^{g_2} = 0.$$

The coefficients in (16) are analytic in $\mathfrak{B}_1$ of §3. In some area in $\mathfrak{B}_1$, (16) has $g_2$ analytic solutions. Let $c_2(x)$ be one of these. We assign an area $\mathfrak{B}_2$ to $c_2$ in such a way that if $c_2$ is of multiplicity $s_2$, the $s_2^{\text{th}}$ derivative with respect to c of the first member of (16) is distinct from zero throughout $\mathfrak{B}_2$ for $c = c_2(x)$. We understand that $\mathfrak{B}_2$ lies with its boundary inside of $\mathfrak{B}_1$.

It may be that $c_2\, y^{p_2}$ annuls $F_1$ when substituted for $z_1$. In that case

$$c_1\, y^{p_1} + c_2\, y^{p_2}$$

is a series (4) which annuls F. If $c_2 y^{p_2}$ does not annul $F_1$, we put $z_1 = c_2 y^{p_2} + z_2$, we write

$$F_1(x, y, z_1) = F_2(x, y, z_2),$$

and we give $F_2$ the treatment accorded to F and $F_1$.

7. It may be that at some stage in our process we are led to a finite series.

$$c_1(x) y^{p_1} + \ldots + c_r(x) y^{p_r}$$

which annuls F. In what follows, we assume that this does not happen, so that we are led to consider an infinite sequence of terms

(17) $\qquad c_1 y^{p_1}, \ldots, c_r y^{p_r}, \ldots$

where the p increase with their subscripts and the c are all analytic at some point $\underline{a}$ contained in $\mathfrak{B}_1, \mathfrak{B}_2, \ldots$.

8. We shall be able to build a series (4) out of (17) through the proof of various facts. We shall show that the p have a common denominator. The $c(x)$, analytic at $\underline{a}$, will be seen to be continuable over an area $\mathfrak{C}$ containing $\underline{a}$. The series

(18) $\qquad c_1(x) y^{p_1} + \ldots + c_r(x) y^{p_r} + \ldots$

will be found to converge for x in $\mathfrak{C}$ and for $|y|$ small and not zero. If $\underline{s}$ is the common denominator of the $p_i$, the replacement of $y^{1/s}$ by v in (18) will produce a function of x and v, analytic for x in $\mathfrak{C}$ and for $|v|$ small and not zero.

9. Towards proving the above statements, we shall show that $g_2$, the degree of (16), does not exceed the multiplicity $s_1$ of $c_1$. This will show that $g_2 \leq g_1$.

We inspect (15). The term $(\sigma_0 - \sigma_i')/i$ is a maximum for $i = s_1$. The term $(\sigma_0' - \sigma_0)/i$ is less for $i > s_1$ than for $i \leq s_1$. Hence the greatest value of i for which the first member of (15) is a maximum cannot exceed $s_1$. Then $g_2 \leq s_1 \leq g_1$.

10. We can now prove that the p have a common denominator. From §9 we see that for j large, say for $j > e$, where e is some integer, the $g_j$ have a common value, say q. Let j have any fixed value greater than e. The equation

(19) $\qquad k_0^{(j)} + \ldots + k_q^{(j)} c^q = 0$

which determines $c_j$, must have q equal solutions. The first member of (19) is therefore of the form

(20) $$k_q^{(j)} (c - c_j)^q.$$

This means, because $c_j \neq 0$, that $k_1^{(j)}$ in (19) is not zero. It follows that the ratio

$$\frac{\sigma_0^{(j-1)} - \sigma_i^{(j-1)}}{i}$$

attains its maximum value $p_j$ for $i = 1$. This means that we can use, for the denominator of $p_j$, the common denominator of the exponents of y in $F_{j-1}(x, y, z_{j-1})$. The same denominator can be used for the $p_i$ with $i > j$.

It follows that the $p_i$ increase towards $+\infty$ with $i$.

11. We shall show that the series (18) is a *formal* solution of $F(x, y, z) = 0$. This will help us to establish analyticity properties of (18).

By §5, the quantity $\sigma_0^{(j)}$ increases with $j$. As the $\sigma_0^{(j)}$ all have a common denominator, namely, that of the p, it must be that $\sigma_0^{(j)}$ increases towards $+\infty$ with $j$. Now the substitution

(21) $$z = c_1 y^{p_1} + \ldots + c_j y^{p_j} + z_j$$

converts $F(x, y, z)$ into

(22) $$F_j(x, y, z_j) = z_j^{(n)} + \ldots + B_0^{(j)}.$$

In (22), $B_0^{(j)}$ is the result of substituting $c_1 y^{p_1} + \ldots + c_j y^{p_j}$ for $z$ in $F(x, y, z)$. The lowest exponent of y in $B_0^{(j)}$ is $\sigma_0^{(j)}$, which is large if $j$ is large. It follows that (18) satisfies (2) formally.

12. Moving towards the completion of our investigation of (18), we shall prove that $q$ in (19) is unity. Let $q > 1$. For $j$ large we have, by §10,

$$p_{j+1} = \sigma_0^{(j)} - \sigma_1^{(j)} = \frac{\sigma_0^{(j)} - \sigma_q^{(j)}}{q}.$$

Hence

(23) $$\sigma_1^{(j)} = \sigma_q^{(j)} + (q - 1) p_{j+1}.$$

It is easy to see from (2), (21), and (22) that the set of all numbers $\sigma_i^{(j)}$, for all $i$ and $j$, is bounded from below. Indeed, if o is the least of the $o_i$ associated with (2), no $\sigma_i^{(j)}$ is less than

$\sigma - n |p_1|$. From (23) it follows, because $q > 1$, that $\sigma_1^{(j)}$ tends towards $+\infty$ with $j$. Now, by (22),

$$(24) \qquad \frac{\partial F_j(x, y, z_j)}{\partial z_j} = n z_j^{(n-1)} + \ldots + 2 B_2^{(j)} z_j + B_1^{(j)}.$$

The least exponent of $y$ in $B_1^{(j)}$ is $\sigma_1^{(j)}$. Also $B_1^{(j)}$ is the result of replacing $z$ in $\partial F(x, y, z)/\partial z$ by $c_1 y^{p_1} + \ldots + c_j y^{p_j}$. It follows that (18) is a formal solution of $\partial F/\partial z = 0$. This contradicts the fact that (2) is irreducible. Thus $q = 1$.

13. We have, analogously to (11),

$$(25) \qquad F_j(x, y, z_j) = \varphi_j(x, c_j + z_j y^{-p_j}) y^{\sigma_0^{(j-1)}} + \psi_j(x, y, c_j + z_j y^{-p_j}) y^{\tau^{(j-1)}}$$

with $\tau^{(j-1)} > \sigma_0^{(j-1)}$. For $j > e$, we have, by (20),

$$(26) \qquad \varphi_j(x, c) = k_1^{(j)}(x) [c - c_j(x)].$$

We put, in (25), $z_j = y^{p_j} u$. Then

$$c_j + z_j y^{-p_j} = c_j + u$$

and the equation $F_j(x, y, z_j) = 0$ gives, for $u$, the equation

$$(27) \qquad k_1^{(j)}(x) u = - \psi_j(x, y, c_j + u) y^{\tau^{(j-1)} - \sigma_0^{(j-1)}}.$$

Let $s$ be the common denominator of the $p$ in (17). We put $y = v^s$ in (27), and (27) becomes an equation for $u$ in terms of $x$ and $v$,

$$(28) \qquad k_1^{(j)} u = D_0(x, v) + D_1(x, v) u + \ldots + D_n(x, v) u^n.$$

The $D$ are series of positive integral powers of $v$ with coefficients analytic in $\mathfrak{B}_j$ which represent functions analytic for $x$ in $\mathfrak{B}_j$ and for $|v|$ small. From (26) we see that $\mathfrak{B}_j$ is taken so that $k_1^{(j)}(a) \neq 0$, where $\underline{a}$ is as in §7.

We are now able to solve for $u$ in (28) by means of the implicit function theorem. As $D_i(a, 0) = 0$ for every $i$ and as $k_1^{(j)}(a) \neq 0$, (28) defines $u$ as a function of $x$ and $v$, analytic for $x$ in some area $\mathfrak{C}$ containing $\underline{a}$, and for $|v|$ small.

Now $F_j(x, y, z_j)$ is annulled formally when $z_j$ is replaced by the series

(29) $$c_{j+1} y^{p_{j+1}} + \dots .$$

This means that (28) is satisfied by the series obtained by dividing the series in (29) by $y^{p_j}$ and then replacing $y^{1/s}$ by $v$. But (28) is satisfied by just one series of positive powers of $v$ with coefficients analytic at $a$, namely, the expansion of the solution of (28) found above. Thus the coefficients in (18) are continuable over $\mathfrak{C}$, and (18) converges for $x$ in $\mathfrak{C}$ and for $|y|$ small but not zero. From this follow the properties of (18) stated in §8.

14. We have obtained one of the series (4); let it be called $P_1$. By division, we find

$$F(x, y, z) = (z - P_1) G(x, y, z)$$

where G is degree n-1 in z, with coefficients of the type of the B in (2), except that y may enter in fractional powers. If $n > 1$, we can treat G as F was treated, to obtain a second series (4). The fractional powers of y constitute no difficulty. We cannot use irreducibility for G as we did for F in §12. However, if $q > 1$ for G, G will have a multiple formal solution, so that the same will be true for F.

The proof of the existence of the n series (4) is thus completed.

15. Let us suppose now that the $A_i$ in (1) are polynomials in y. We are interested in the neighborhood of $y = \infty$. Dividing by $A_n$, we obtain an equation (2) with each $B_i$ a series of *descending* integral powers of y. Using a transformation $y = 1/t$, we show the existence of n series (4) where now the $p_i$ *decrease* as their subscripts increase.

The $a_{ji}(x)$ in (3) are rational combinations of the coefficients of the A. The function $c_1(x)$ in (4), determined by (6), is algebraic in a certain number of the $a_{ji}$. Again, $c_2$, determined by (16), is algebraic in $c_1$ and some of the $a_{ji}$. In this way we find that *every $c(x)$ in (4) is algebraic in the coefficients of the $A(x, y)$.*

## Chapter VI

## INTEGRATION OF DIFFERENTIAL EQUATIONS BY QUADRATURES

### INTEGRABILITY BY QUADRATURES

1. In the formal theory of differential equations, one meets frequently equations whose solutions contain integrals which are not elementary functions. For instance, the equation

$$x y' + x y = 1$$

has the solution

$$y = e^{-x} \int \frac{e^x \, dx}{x} + c \, e^{-x},$$

involving an integral which is not elementary. It is customary to regard an equation as solved when a solution of the above general type can be found for it. In addition to algebraic functions, exponentials, and logarithms, one admits the operation of integration. When for a differential equation the unknown function can be expressed, naturally with arbitrary constants, by means of the operations just indicated, the equation is said to be *integrable by quadratures*. Also, giving a new meaning to the word *terms*, one describes the equation as integrable in finite terms. A *quadrature*, of course, is simply an integration.

Since $\log f(x)$, for any $f(x)$, is an integral of $f'/f$, we may dispense with the operation of taking a logarithm and limit ourselves to algebraic operations, exponentiations, and integrations.

We are dealing now with the representation of the unknown function in explicit form. In the chapters which follow, we shall consider problems of implicit representation.

It is our intention to present the results of an investigation of Liouville on the cases of integrability of a type of Riccati equation and of Bessel's equation. For this we first construct, with the operations indicated above, a category of functions.

### FUNCTIONS OF LIOUVILLE

2. The variable x will be called a complete monomial of order 0 and every algebraic function of x will be called a function of order 0.

An m.a.f. is called a complete monomial of order 1 if it is not algebraic and if it is either the exponential of an algebraic function or the *integral* of an algebraic function. As has already been observed, a logarithm of an algebraic function is also an integral of an algebraic function. A branch of a complete monomial is called a monomial.

One continues as in Chapter I. The functions of any order n are fluent functions. Their structure is described as in I, 11, integrals replacing logarithms.

The functions to which orders have just been assigned will be called *functions of Liouville*.

### EQUATIONS OF RICCATI AND BESSEL

3. The differential equation

$$y' = P(x) + Q(x) y + R(x) y^2$$

is known as the equation of Riccati. Daniel Bernouilli studied the special Riccati equation

(1) $$y' + y^2 = x^n$$

with n a constant. He found the equation to be integrable by quadratures when $n = -2$ or when $n = -4p/(1 + 2p)$ with p an integer, positive, negative, or zero. Liouville [7], [9], and [23], showed that these are the only cases of integrability by quadratures. In perfectly definite language, these are the only cases in which (1) has even a single solution which is a function of Liouville.

Equation (1) is intimately related to the equation of Bessel

(2) $$x^2 y'' + x y' + (x^2 - v^2) y = 0,$$

with ν constant, which, as will be seen, admits a Liouville function distinct from zero as a solution when and only when 2ν is an odd integer.

### A THEOREM OF LIOUVILLE

4. The proofs of the results stated in §3 are based on the following theorem.

THEOREM: *If the equation*

(3) $$y' + y^2 = P(x)$$

*with P algebraic, has a special solution which is a Liouville function, it has a special solution which is algebraic.*

In the proof, there will appear the procedure to which reference was made in Chapter V.

One should have well in mind what it means for a function $y(x)$ to satisfy (3). What is meant is that some pair of elements of $y$ and of $P$, with the same center, satisfy (3).

When $P = 0$, (3) has the solution $1/x$. In what follows, we assume that $P$ is not identically zero.

Let us suppose that (3) is satisfied by Liouville functions but no algebraic function. Let E denote the totality of Liouville functions satisfying (3). Those functions in E which are of a least order m form a class $E_1$. Of course $m > 0$. We use for each function in $E_1$ an expression involving as few m-monomials as possible. Let $y(x)$ be a function in $E_1$ involving no more m-monomials than appear in any other function in $E_1$.

Now let $\theta$ be one of the m-monomials in $y$. We write $y = F(\theta, x)$. By (3),

(4) $\qquad F_\theta(\theta, x)\, \theta' + F_x(\theta, x) + [F(\theta, x)]^2 = P(x).$

If $\theta$ is an exponential, $e^v$, (4) becomes

(5) $\qquad\qquad F_\theta\, \theta\, v' + F_x + F^2 = P.$

If $\theta$ is the integral of a function $v$,

(6) $\qquad\qquad F_\theta\, v + F_x + F^2 = P.$

A relation (5) or (6) must be an identity in $\theta$ and $x$. What we are going to do is to expand the first members of (5) and (6) in descending powers of $\theta$, using the results given in V, 15. The question of the subsistence of the relations (5) and (6) for values of the independent variable $\theta$ in the neighborhood of $\infty$ will be examined later.

The function $F(\theta, x)$, with $\theta$ an independent variable, is defined by an equation like (1) of Chapter V. The place of $\theta$ is taken in that equation by $y$. The coefficients in the A are polynomials in the monomials other than $\theta$ which enter into F. We have thus

(7) $\qquad F(\theta, x) = c_1(x)\, \theta^{p_1} + c_2(x)\, \theta^{p_2} + \ldots$

where the p decrease as their subscripts increase. Each c, by V, 15, is algebraic in the monomials other than $\theta$ which appear in F. The first member of (5) becomes

(8) $\qquad [(c_1' + p_1\, c_1\, v')\, \theta^{p_1} + \ldots] + [c_1^2\, \theta^{2p_1} + \ldots]$

while that of (6) becomes

(9) $\qquad [c_1' \theta^{p_1} + \ldots] + [c_1^2 \theta^{2p_1} + \ldots].$

As P is not zero, the series obtained by simplifying that one of (8) and (9) which is pertinent must start with a term of exponent 0. It follows that, whether $\theta$ is an exponential or an integral, $p_1 \geq 0$. If $p_1 > 0$, there will be, both in (8) and in (9), a term in $\theta^{2p_1}$ which cannot cancel. Hence $p_1 = 0$, and we obtain from either (5) or (6),

$$c_1' + c_1^2 = P.$$

Thus $c_1$ is a solution of (3). Now $c_1$ is either of order less than m or else involves fewer m-monomials than $F(\theta, x)$ does. We have here a contradiction which proves the theorem of Liouville.

5. We have now to justify the use of the series (7). Let $B$ be an area in the plane of x for which $F(\theta, x)$ has a set of expansions (7), the number of expansions being the degree of the equation, analogous to (1) of V, 1, which defines $F(\theta, x)$. Let $K(\theta, x)$ be, for that equation, the product of the discriminant by the coefficient of the highest power of $F(\theta, x)$. We work at a point $\underline{a}$, lying in $B$, for which

$$K(\theta(a), a) \neq 0.$$

Now K is a polynomial in $\theta$, and thus the function $K(\theta, a)$ of $\theta$ can vanish for at most a finite number of values of $\theta$. Suppose now that we have a curve C, in the space of x and $\theta$, defined by equations

$$x = a, \quad \theta = \varphi(t), \quad 0 \leq t \leq 1,$$

with $\varphi(0) = \theta(a)$, the function $\varphi(t)$ having a large modulus for $t = 1$ and assuming nowhere on $(0, 1)$ any of the values of $\theta$ which annul $K(\theta, a)$. Then $F(\theta, x)$ can be continued along C from $t = 0$ to $t = 1$. As $\varphi(1)$ has a large modulus, some expansion (7) represents $F(\theta, x)$ in the neighborhood of the terminal point of C, while the first members of (5) and (6) are represented by (8) and (9). The validity of our method is thus established.

The foregoing method was presented by the author [20] in 1926. In some cases, as in that above, it works with surprising speed. A trace of the idea involved appears to exist in Mordukhai-Boltovskoi's book on integration ([13], p. 197), where, in connection with certain rational combinations of a monomial $\theta$, expansions in descending integral powers of $\theta$ are used.

Liouville ([4], p. 65, and [8], p. 442) refers to expansions of algebraic functions in ascending or descending powers. He states that he originally used such series in certain proofs but later eliminated them from his work. The proofs to which he refers are proofs, not of general theorems, but rather of results on special functions, obtained as applications of general theorems. He appears to have in mind expansions of algebraic functions of one variable and, perhaps, proofs like that of II, 13. The diffidence towards infinite series is understandable; it was a diffident decade. All that was to disappear soon under the leadership of Cauchy and Weierstrass. It is hard to guess at the types of expansions of algebraic functions which he used, since the general theory of such expansions dates only from Puiseux's investigation of 1854.

### APPLICATION TO EQUATIONS OF RICCATI AND BESSEL

6. We take up now Riccati's equation (1) and Bessel's equation (2). We put $y = x^{-1/2} u$, $x = i z$, and (2) goes over into

(10) $$\frac{d^2 u}{dz^2} = \left(1 + \frac{p(p+1)}{z^2}\right) u$$

where $p = \nu - 1/2$. Putting $v = u'/u$, we have

(11) $$\frac{dv}{dz} + v^2 = 1 + \frac{p(p+1)}{z^2}.$$

We now consider (1). For $n = -2$, the substitution $y = v/x$ renders (1) separable, and the solutions of (1) are seen to be elementary functions. In what follows, we suppose that $n \neq -2$. We put $y = w'/w$. Then (1) gives

(12) $$w'' = x^n w.$$

Let $n = 2q - 2$. Then $q \neq 0$. We put $z = x^q/q$ and (12) becomes

(13) $$\frac{d^2 w}{dz^2} + \frac{q-1}{q} \frac{1}{z} \frac{dw}{dz} - w = 0.$$

We let $p = (1-q)/(2q)$ and put $w = z^p u$, whereupon (13) goes over into (10).

7. The question of the satisfaction of (1) by a function of Liouville or of (2) by such a function other than 0 reduces thus to the question of the existence of a Liouville function satisfying (11). By §4, we must seek those values of $p$ for which (11) has an algebraic solution. We shall show that an algebraic solution exists when and only when $p$ is an integer. This will validate the statements of §2.

We show first that every algebraic solution of (11) is rational. An algebraic function of z which is not rational must have at least two critical points. That is, for at least two values of z, the Riemann surface of the function must have branch points. At a branch point, a function has an expansion in which fractional powers are effectively present.

We show first that no algebraic solution v of (11) can have a branch point at $\infty$. Let v have a branch point at $\infty$, and an expansion

(14) $$v = a_1 z^{p_1} + a_2 z^{p_2} + \ldots$$

where the p are decreasing fractions. We suppose that no $\underline{a}$ is zero. It is seen from (11) that $p_1 = 0$. Let $p_i$ be the highest fractional exponent. Then the highest fractional power in $v^2$ is found in $2 a_1 a_i z^{p_i}$. But this cannot be cancelled by any term in v', in which the highest fractional power comes from $p_i a_i z^{p_i - 1}$. Thus $p_i$ is not a fraction, and there is no branch point at $\infty$.

We show now that there is no branch point for any finite value $c \neq 0$ of z. Suppose that v has, at c, an expansion

$$v = a_1 (z - c)^{p_1} + \ldots .$$

We shall show first that $p_1$ is not a fraction, then that no $p_i$ is a fraction.

Suppose that $p_1$ is fractional. The first term in v' is $p_1 a_1 (z - c)^{p_1 - 1}$ and has a fractional exponent. The right member of (11) is rational, and hence its development in powers of $z - c$ for any c has only integral exponents. Thus there must be a term in $v^2$ which balances the first term of v'. Hence for some i and j,

$$p_1 - 1 = p_i + p_j .$$

But $p_i + p_j \geq 2 p_1$. Thus $2 p_1 \leq p_1 - 1$ or $p_1 \leq -1$. Now $p_1$ cannot be less than $-1$, else the first term of $v^2$, which is $a_1^2 (z - c)^{2p_1}$, could not be balanced by any term in v' or in the second member of (11). Thus $p_1$ is integral.

Suppose now that some $p_i$ with $i > 1$ is fractional and indeed the least fractional exponent. The least fractional exponents in v' and $v^2$ will be found in

$$p_i a_i (z - c)^{p_i - 1}, \quad 2 a_1 a_i (z - c)^{p_1 + p_i}$$

We must therefore have $p_1 = -1$ and $p_i = -2 a_1$. The terms of lowest degree in v' and $v^2$ are respectively,

## INTEGRATION BY QUADRATURES

$$-a_1(z - c)^{-2}, \quad a_1^2(z - c)^{-2}$$

The expansion of the second member of (11) about $z = c$ has no negative powers, since $c \neq 0$. Then

$$-a_1 + a_1^2 = 0$$

so that $a_1 = 1$. Hence $p_1 = -2$, and $p_1$ is not a fraction.

Thus an irrational algebraic solution of (11) could have a critical point only at $z = 0$. Then every algebraic solution of (11) is rational.

8. We are going to show that, if (11) has a rational solution, $p$ is an integer. Let (11) have a solution

$$v = \frac{P(z)}{Q(z)}$$

with $P$ and $Q$ polynomials with no zero in common. From (11), we see that $v$ has no pole at $\infty$ and that the poles of $v$ in the finite plane are all simple poles. There is certainly a pole at $z = 0$, and if there is a pole at $c \neq 0$, the term in $1/(z - c)$ has unity for coefficient. Let $v(\infty) = h$ and let the zeros of $Q$ be $c_1, \ldots, c_r$. From (11) we see that $h \neq 0$. Then

$$v = h + \frac{k}{z} + \frac{1}{z - c_1} + \ldots + \frac{1}{z - c_r}.$$

We have for $k$ the equation

(15) $$k^2 - k = p(p + 1),$$

so that $k = p + 1$ or $k = -p$. The development of $v$ about $\infty$ is

$$v = h + \frac{k + r}{z} + \ldots .$$

Now $k + r$ must be zero or the term $2h(k + r)z^{-1}$ would be present in $v^2$ and would not be cancelled by $v'$ or the second member of (11). Thus $k$ is an integer. This means, by (15), that $p$ is an integer.

9. We show finally that when $p$ is an integer, (11) has a rational solution. There is no generality lost in assuming that $p$ is positive. When $p$ is either 0 or $-1$, (11) is satisfied by $v = \pm 1$. If $p < -1$, we let $p' = -p - 1$. Then

$$p(p + 1) = p'(p' + 1).$$

We put $v = w'/w$ in (11) so that

$$w'' = \left(1 + \frac{p(p+1)}{z^2}\right) w.$$

Now let $w = e^z z^{-p} u$. Then

(16) $$u'' + 2\left(1 - \frac{p}{z}\right) u' - 2\frac{pu}{z} = 0.$$

Let us determine the power series

$$u = a_0 + a_1 z + \ldots + a_m z^m + \ldots$$

so as to satisfy (16). We find equations

$$-2 p a_1 - 2 p a_0 = 0$$

$$a_2(2 - 4p) + a_1(2 - 2p) = 0$$

. . . . . . . . . . . . . . . . . . . . . .

$$a_m [m(m-1) - 2mp] + a_{m-1}(2m - 2 - 2p) = 0.$$

We are supposing that p is a positive integer. Our equations give $a_m$ in terms of $a_{m-1}$ unless $m - 1 = 2p$. But when m is $p + 1$, which is less than $2p + 1$, we find that $a_m = 0$. Thus, if we take $a_m = 0$ for $m > p + 1$, we get a polynomial u which satisfies (16). The logarithmic derivative of the corresponding w is rational and satisfies (11).

This concludes our treatment of the equations of Riccati and Bessel. The existence of a single solution of (1) of the Liouville type is easily seen to imply that all solutions are of that type. The same holds for (2) if we disregard the solution $y = 0$.

## FURTHER STUDIES

10. Mordukhai-Boltovskoi [12] investigated the integrability by quadratures of linear differential equations of any order.

In the following chapter we shall go with Mordukhai-Boltovskoi into the question of integrating algebraic differential equations of the first order in terms of elementary functions.

There is another type of problem on integrability by quadratures. The linear equation of the first order

(17) $$y' + P(x) y = Q(x),$$

where P and Q are *any* functions, can be integrated by two quadratures. This raises the problem of finding classes of equations, *involving arbitrary functions*, which can be solved by finite algorithms, with integration among the permitted operations. Such a question was examined by Maximovich [11] in 1885. As the author has not been able to secure Maximovich's paper or any account of it except those given in an abstract in the *Jahrbuch* and in one in the Paris *Comptes Rendus*, he is unable to make a definite statement in regard to it. It appears that, with certain assumptions, Maximovich shows that the linear equations are, essentially, the only general class of equations of the first order which can be integrated in explicit form by quadratures.

Chapter VII

# IMPLICIT AND EXPLICIT ELEMENTARY SOLUTIONS OF DIFFERENTIAL EQUATIONS OF THE FIRST ORDER

*IMPLICIT REPRESENTATIONS*

1. In the preceding chapters we have discussed the possibility of representing functions in *explicit* form by means of certain operations. In solving differential equations, one is perfectly happy to end up with a relation among the unknown function, the variable, and arbitrary constants. For instance, the equation

$$\frac{dy}{dx}(1 - e^y) = 1$$

has for solution

$$y - e^y = x + c.$$

For no value of c is y an elementary function of x. There arise thus questions on the possibility of solving differential equations in finite *implicit* terms. We shall consider some problems of this type.

*ELEMENTARY FUNCTIONS OF TWO VARIABLES*

2. For our purposes, we must construct the elementary functions of two variables, x and y. We consider an m.a.f. $u(x, y)$. An element of u is a power series $P(x - x_0, y - y_0)$. By the radius of convergence r of P we shall mean the least upper bound, finite or infinite, of those positive numbers $\rho$ which are such that P converges for $|x - x_0| < \rho$, $|y - y_0| < \rho$. An immediate continuation of P is an element secured by developing P in powers of $x - x_1$, $y - y_1$, where $|x_1 - x_0| < r$, $|y_1 - y_0| < r$.

We shall call u *fluent* if for each element $P(x-x_0, y-y_0)$ of u, for each curve

(1) $\qquad x = \varphi(\lambda), \quad y = \psi(\lambda) \quad (0 \leq \lambda \leq 1)$

where $\varphi(0) = x_0$, $\psi(0) = y_0$, and for each $\varepsilon > 0$, there exists a curve

(2) $\qquad x = \varphi_1(\lambda), \quad y = \psi_1(\lambda) \quad (0 \leq \lambda \leq 1)$

with $\varphi_1(0) = x_0$, $\psi_1(0) = y_0$, such that

(3) $\qquad |\varphi_1(\lambda) - \varphi(\lambda)| < \varepsilon, \qquad |\psi_1(\lambda) - \psi(\lambda)| < \varepsilon$

for $0 \leq \lambda \leq 1$ and such that P can be continued along (2).

In proving the fluency of functions, we shall use the following fact. *Given a* u(x, y), *not identically zero, with an element* $P(x - x_0, y - y_0)$ *continuable along a curve (1), and given any* $\varepsilon > 0$, *there is a curve (2) satisfying (3) such that P is continuable along (2) and that u is nowhere zero on (2), except perhaps at* $(x_0, y_0)$. To prove this we suppose, as we may, that (1) consists of short straight segments with extremities at $(x_0, y_0)$, $(x_1, y_1)$, ..., $(x_n, y_n)$, the point $(x_n, y_n)$ being the terminal point of (1). We may suppose, furthermore that u is not zero at any $(x_i, y_i)$ with $i > 0$. If each $(x_i, y_i)$ with $i > 0$ is close enough to $(x_{i-1}, y_{i-1})$, u will be continuable along (1) by a chain of elements $P(x - x_i, y - y_i)$, each after the first an immediate continuation of its predecessor. The first segment is given by

(4) $\qquad x = x_0 + t(x_1 - x_0), \qquad y = y_0 + t(y_1 - y_0), \qquad 0 \leq t \leq 1$.

If x and y are replaced in $P(x - x_0, y - y_0)$ by their expressions in (4), P becomes a function Q(t) analytic for $|t| \leq 1$ and not zero for $t = 1$. Let $t = 0$ be joined to $t = 1$ by a curve $\tau = \xi(t), 0 \leq t \leq 1$ with $|\tau - t|$ very small along the curve, Q being distinct from 0 along the curve, except perhaps at the first point. Then $P(x - x_0, y - y_0)$ is analytic along

(5) $\qquad x = x_0 + \xi(t)(x_1 - x_0), \qquad y = y_0 + \xi(t)(y_1 - y_0) \qquad 0 \leq t \leq 1$,

and will not be zero on (5) except perhaps at the first point. One makes similar replacements for the other segments.

3. The variables x and y will be called complete monomials of order zero. An algebraic function u(x, y) will be called a function of order zero. An algebraic u(x, y) is fluent, for, given any curve (1), we can replace it by a curve (2) close to it along which, except perhaps at the first point, the discriminant of the equation defining u and the coefficient of the highest power of u are distinct from zero. If u is algebraic and nonconstant, $e^u$ and log u are called complete monomials of order 0. Of course $e^u$ is fluent; so also, by §2, is log u. A branch of a complete monomial, analytic in a region in the space of x and y, is called a monomial. Continuing, we obtain the *elementary functions of x and y*. The structure of such a function is described as in I, 11. One uses, in the

description, a branch of a function u of order n, analytic in a region $\mathfrak{U}$. One employs, in the place of $\underline{a}$ of I, 11, a point (a, b) in $\mathfrak{U}$. In (I), one uses $r_1$ algebraic functions of x and y, analytic for $|x - a| < \rho_1$, $|y - b| < \rho_1$. In (N + 1), one uses an algebraic function of x, y, $x_i'$, ..., $x_{r_n}^{(n)}$. If a function has the structure just described at (a, b), it is said to be of *regular structure* at (a, b).

## IMPLICIT SOLUTIONS OF ALGEBRAIC DIFFERENTIAL EQUATIONS

4. Mordukhai-Boltovskoi [14] investigated differential equations of the form

(6) $$y' = f(x, y)$$

with f algebraic in both variables, inquiring as to whether (1) has a solution $g(x, y) = c$ with c an arbitrary constant and g elementary. He proved the following theorem.

**THEOREM:** *Let* $y' = f(x, y)$ *with* f *algebraic in* x *and* y *have a solution* $g(x, y) = c$ *with* g *elementary. Then it has a solution*

(7) $$\varphi_0(x, y) + a_1 \log \varphi_1(x, y) + \ldots + a_r \log \varphi_r(x, y) = c$$

*with each* $a_i$ *a constant and each* $\varphi_i$ *algebraic.*

The form which the $\varphi$ can be given is discussed in [14].

To prove this theorem, we start by observing that for $g(x, y) = c$, where g is not a constant, to be a solution of (6), it is necessary and sufficient that

(8) $$\frac{\partial g}{\partial x} + f \frac{\partial g}{\partial y} = 0$$

identically in x and y.

Let us suppose that we have a nonconstant elementary solution g of (8) whose order n is as small as possible and which involves as few as possible, say r, n-monomials. We need only consider values of n which exceed 0. Let $\theta$ be one of the n-monomials. We write

$$g(x, y) = h(\theta, x, y).$$

Then

(9) $$h_\theta \theta_x + h_x + f(h_\theta \theta_y + h_y) = 0.$$

5. Now, let $\theta$ be a logarithm, log v. Then (8) becomes

(10) $\qquad h_\theta v^{-1} v_x + h_x + f(h_\theta v^{-1} v_y + h_y) = 0$

and (10) holds identically in $\theta$, $x$, $y$.

Let one of the expansions of $h(\theta, x, y)$ for $\theta = \infty$ be

(11) $\qquad\qquad \alpha_1 \theta^{p_1} + \alpha_2 \theta^{p_2} + \ldots,$

the $\alpha$ being algebraic in the monomials other than $\theta$ in $g$. We assume that $\alpha_1 \neq 0$. We find from (10) that

(12) $\qquad\qquad (\frac{\partial \alpha_1}{\partial x} + f \frac{\partial \alpha_1}{\partial y}) \theta^{p_1} + \ldots = 0$

identically in $x$, $y$, $\theta$ for $|\theta|$ large. The coefficient of $\theta^{p_1}$ in (12) is zero. This means that $\alpha_1$ is a constant. Otherwise $\alpha_1$ would be a solution of (8) involving fewer than r-monomials of order n. Let $\alpha_1 = a$. If now $p_1 = 0$, we replace $g$ by $g - a$ and use the expansion $\alpha_2 \theta^{p_2} + \ldots$, with $p_2 < 0$ and $\alpha_2 \neq 0$. Then $\alpha_2$ is a constant. We therefore make the legitimate assumption that $p_1 \neq 0$. Let $p_1 \gamma$ represent the coefficient of $\theta^{p_1-1}$ in (11). The coefficient of $\theta^{p_1-1}$ (12) is

$$(p_1 a v^{-1} v_x + p_1 \gamma_x) + f(p_1 a v^{-1} v_y + p_1 \gamma_y)$$

and that coefficient is zero. Hence $a \theta + \gamma$ is a solution of (8). It is not a constant; otherwise $\theta$ would be expressible in the other monomials. Suppose now that $\gamma$, whose monomials are among those of $g$, involves a second logarithm, $\zeta = \log w$, of order n. We develop $\gamma$ with respect to $\zeta$ at $\zeta = \infty$. Then

$$a \cdot \theta(x, y) + \gamma = \beta_1 \zeta^{p_1} + \ldots.$$

We see as above that $\beta_1$ is a constant $b$, that we may assume that $p_1 \neq 0$, and that $b \zeta + \delta$, with $p_1 \delta$ the coefficient of $\theta^{p_1-1}$, is a solution of (8). Now if $p_1 - 1$ is not zero, $\delta$ will not involve $\theta$, and $b\zeta + \delta$ will be too simple a solution of (8). Thus $p_1 = 1$ and $\delta = a \theta + \rho$ with $\rho$ free of $\theta$ and $\zeta$. Then $a \theta + b \zeta + \rho$ is a solution of (8). Continuing, we find that (8) has a solution

(13) $\qquad\qquad a_1 \theta_1 + \ldots + a_s \theta_s + \tau$

where the $\theta$ are n-logarithms and $\tau$ involves no n-logarithm.

We shall show that no n-exponential appears in $\tau$. Let there be such an exponential, $\theta = e^v$. Let the expression (13) have an expansion given by (11). We find an identity

$$\text{(14)} \quad \sum_{i=1}^{\infty} \left[\frac{\partial \alpha_i}{\partial x} + p_i \alpha_i v_x + f\left(\frac{\partial \alpha_i}{\partial y} + p_i \alpha_i v_y\right)\right] \theta^{p_i} = 0.$$

Let $\gamma$ be the coefficient of $\theta^0$ in (11). We see from (13) that

$$\text{(15)} \quad \gamma = a_1 \theta_1 + \ldots + a_s \theta_s + \xi$$

where $\xi$ is free of $\theta_1, \ldots, \theta_s, \theta$. Then (14) gives

$$\gamma_x + f \gamma_y = 0$$

and we have in $\gamma$ a solution of (8) with fewer than $r$ monomials of order $n$. Thus, if $g$ contains an n-logarithm, (8) has a solution of the form (7) with each $\log \varphi_i$ an n-logarithm and with $\varphi_0$ of order less than $n$.

6. Suppose now that $g$ contains no n-logarithm. Let $\theta = e^v$ be an n-exponential in $g$. We use an expansion (11) for $g$ and find the identity (14). Consider some $i$ for which $p_i$ and $\alpha_i$ are both distinct from zero. We equate to zero the coefficient of $\theta^{p_i}$ in (14). We multiply the resulting relation through by $\theta^{p_i}$, where $\theta = e^v$ and integrate. We find that $\alpha_i \theta^{p_i}$ is a solution of (8). It is not a constant; otherwise $\theta$ would be algebraic in the other monomials in $g$. We can see now that there is no other exponential among the n-monomials in $g$. If there were such a monomial, $\zeta = e^w$, we would obtain from $\alpha_i \theta^{p_i}$ a solution of (8)

$$\text{(16)} \quad \beta_j(x, y) \zeta^{q_j} \theta^{p_i}$$

with $\beta_j$ free of $\zeta$ and $\theta$. We could write (16)

$$\beta_j e^{q_j w + p_i v}$$

and (8) would have a solution involving fewer than $r$ n-monomials. Thus Thus $\alpha_i$ is of order less than $n$. Also

$$\text{(17)} \quad \log \alpha_i \theta^{p_i} = p_i v + \log \alpha_i$$

is a solution of (8). Then $\log \alpha_i$ must be of order $n$, and we have in (17) a solution of (8) of the type of the first member of (7).

7. We shall now prove that $n = 1$. The integer $r$ being as in §4, there are expressions of the form

$$\text{(18)} \quad \varphi_0 + a_1 \log \varphi_1 + \ldots + a_r \log \varphi_r,$$

representing functions of order $n$, which satisfy (8). Each $\varphi_i$ with $i > 0$ is of order $n - 1$ and its logarithm is of order $n$; $\varphi_0$ is of order less than $n$. With each expression we associate two numbers, first s, the number of $(n-1)$-monomials appearing in at least one of $\varphi_0$, $\varphi_1$, ..., $\varphi_r$, then t, the number of such monomials in at least one of $\varphi_1$, ..., $\varphi_r$. We consider those expressions for which s is a minimum and select from them one, $g(x, y)$, for which t is a minimum.

We might use the power series method, but it would be somewhat cumbersome in the present connection. Instead, letting $\theta = e^v$ be one of the $(n-1)$-monomials, we write

$$g(x, y) = h(\theta, x, y).$$

It is easy to see that $h(\mu\theta, x, y)$ satisfies (8) for $\mu$ constant. Also, because f in (8) is free of $\mu$, $h_\mu(\mu\theta, x, y)$ satisfies (8). Then $\theta h_\theta(\theta, x, y)$ satisfies (8). As it is of order less than n, it is a constant. Then it must be a constant when $\theta$ is an independent variable. Thus

$$h(\theta, x, y) = a \log \theta + \gamma(x, y)$$

and, in the usual way, we get an expression for g free of $\theta$. Now, let $\theta = \log v$ be one of the $(n-1)$-monomials in $\varphi_1$, ..., $\varphi_r$. Then $h(\theta + \mu, x, y)$ is a solution of (8). Thus $h_\theta(\theta, x, y)$ satisfies (8) and is a constant. Then, for $\theta$ independent,

$$h(\theta, x, y) = a\theta + \gamma(x)$$

and we find an expression for g in which the n-logarithms are free of $\theta$. This completes the proof of Mordukhai-Boltovskoi's theorem.

## EXPLICIT ELEMENTARY SOLUTIONS

8. Applying what precedes, we shall obtain a theorem of Mordukhai-Boltovskoi on algebraic differential equations with solutions possessing *explicit* elementary expressions.

We consider an equation

(19) $$F(x, y, y') = 0$$

with F a polynomial in x, y, y', irreducible in the field of complex constants. Suppose that (19) admits a solution $y(x)$ which is elementary but not algebraic. Such a $y(x)$ cannot be a singular solution of (19), so that (19) can be converted into the form

(20)' $$y' = f(x, y)$$

with f analytic at some point $(a, y(a))$.

Now let $y = h(\theta, x)$ where $\theta$ is one of the highest monomials in y. We find without difficulty that $h((1 + c) \theta, x)$ or $h(\theta + c, x)$ is a solution of (20) for $|c|$ small according as $\theta$ is an exponential or a logarithm. In either case, $h_c$ cannot vanish for every x when $c = 0$. If it did, it would vanish identically in $\theta$ and x, and h would be free of $\theta$. The equations

$$y = h((1 + c) \theta, x), \quad y = h(\theta + c, x)$$

can therefore be solved in the neighborhood of some point $(a, y(a))$. We obtain a relation

$$g(x, y) = c$$

with g analytic at $(a, y(a))$. Of course, g satisfies (8). We see that g is algebraic in y and in monomials in x.

9. We consider functions $g(x, y)$ algebraic in y and in monomials in x, which satisfy (8). We do not ask that g be of regular structure, or even analytic, at some point $(a, y(a))$. Whatever pertains to the given solution $y(x)$ of (20) will be taken care of by special considerations.

We assume each g to be expressed in such a way that n, the highest of the orders of the monomials in x which it involves, is as low as possible. We call n the x-order of g.

We now consider a solution g of (8), as in the foregoing, of as low an x-order n as is possible. Then n cannot be 0. If it were, $g(x, y)$ would be algebraic in x and y. Since the singularities of g satisfy an equation algebraic in x and y, and since $y(x)$ as in §8 is not algebraic, g can be continued to some point $(a, y(a))$. Then $y(x)$ would reduce g to a constant when substituted for y, and thus $y(x)$ would be algebraic. Then $n > 0$.

10. We now review the argument of §§5 and 6, taking g algebraic in y and of as low an x-order $n > 0$ as it can be with this condition. We are led to one of the following two cases:

Case A. Equation (8) has a solution

(21) $$\varphi_0 + a_1 \log \varphi_1 + \ldots + a_r \log \varphi_r$$

with each $\log \varphi_i$ an n-logarithm in x and with $\varphi_0$ algebraic in y and of x-order less than n.

Case B. Equation (8) has a solution

(22) $$\alpha e^v$$

with $e^v$ an n-exponential in x and $\alpha$ algebraic in y and of x-order less than n.

We consider Case A. We review §7 and find that $n = 1$, so that the $\varphi_i$ are algebraic. As above, we can continue the $\varphi_i$ to points $(a, y(a))$. What is more, because the given $y(x)$ is not algebraic, $\partial\varphi_0/\partial y$ cannot vanish at every $(a, y(a))$. We equate the expression (21) to a constant c and solve for y in terms of x and c. We find a one-parameter family of solutions of (19),

$$y = G(x, a_1 \log \varphi_1 + \ldots + a_r \log \varphi_r - c)$$

where the $\varphi_i$ are algebraic functions of x and G is algebraic in its two arguments. For some c, we get the given solution $\acute{y}(x)$.

We now examine Case B. If $n = 1$, we equate $\alpha e^v$ in (22) to a constant c and solve for y. We find that

$$y = G(c\, e^{-v(x)})$$

where v and G are algebraic. In what follows, we assume that $n > 1$.

In each expression $\alpha e^v$ which satisfies (8), a certain number of $(n-1)$-monomials in x appear in at least one of $\alpha$ and v. We choose an expression in which this number is a minimum, say s. The function

(23) $\qquad v + \log \alpha,$

which we represent by $g(x, y)$, satisfies (8). Let $\theta$ be one of the $(n-1)$-monomials. We write $g(x, y) = h(\theta, x, y)$. Let $\theta$ be an exponential, $e^w$. Then $h(\mu\theta, x, y)$ satisfies (8). So does $\theta\, h_\theta(\theta, x, y)$. The latter solution is algebraic in y and of x-order less than n. Hence it is constant and, indeed, identically in $\theta$. If then

$$v = k(\theta, x), \quad \alpha = k_1(\theta, x, y),$$

we have

$$g(x, y) = b w + k(z_0, x) + \log k_1(z_0, x, y) - b \log z_0$$

with b constant. Then

(24) $\qquad e^g = k_1(z_0, x, y)\, e^{k(z_0, x) + bw - b \log z_0}.$

As $k_1(z_0, x, y)$ is algebraic in y and of x-order less than n, the exponential in (24) must be of x-order n. As there are fewer $(n-1)$-monomials in (24) than appear in $\alpha$ and v, it follows that $\theta$ cannot be an exponential.

It is now easy to show that $v + \log \alpha$ can be written in the form

$$\log \psi(x, y) + \varphi_0(x) + c_1 \log \varphi_1(x) + \ldots + c_s \log \varphi_s(x)$$

where each $\log \varphi_i$ is an $(n-1)$-logarithm; $\varphi_0$ a function of order less than $n-1$; $\psi$ algebraic in y and of x-order less than $n-1$. Then one proves that $n = 2$, so that $\psi$ and the $\varphi$ are algebraic.

Summarizing the results of Cases A and B, we have the theorem of Mordukhai-Boltovskoĭ [15].

**THEOREM:** *If an algebraic differential equation $F(x, y, y') = 0$ has a special solution which is an elementary, but not algebraic, function of x, the equation has either a one-parameter family of solutions of the type*

$$y = G(x, a_1 \log \varphi_1(x) + \ldots + a_r \log \varphi_r(x) - c)$$

*with c an arbitrary constant, the a constants, and G and the $\varphi$ algebraic, or the equation has a one-parameter family of solutions*

$$y = G(x, e^{\varphi_0(x)} + a_1 \log \varphi_1(x) + \ldots + a_r \log \varphi_r(x) - c)$$

*of similar description.*\*

The question of testing for the elementary integrability of $F = 0$ is considered in [15]. This question connects, on the one hand, with investigations of Painlevé on the singularities of solutions of differential equations and, on the other, with fragmentary researches of Darboux, Poincaré, Painleve, and D'Autonne on algebraic differential equations with algebraic solutions.

\* Note that $\varphi_0$ is also algebraic.

## Chapter VIII
## FURTHER IMPLICIT PROBLEMS

### INTEGRALS OF ELEMENTARY FUNCTIONS

1. Let y(x) be an elementary function of x and let w be an integral of y. We propose to examine the circumstances under which there exists a relation

(1) $$F(w, x) = 0$$

with F elementary in w and x and not identically zero. We shall prove that, when such a relation exists, w(x) is elementary [18]. For instance, if the inverse of an integral of an elementary function is elementary, the integral satisfies an equation (1). Thus, *if the inverse of an integral of an elementary function is elementary, the integral is itself elementary.* For instance, the elliptic function which is the inverse of the integral of the first kind in II, 13, is not elementary. As any two elliptic functions with the same period parallelogram are algebraically related, it follows that no nonconstant elliptic function is elementary.

### FORMULATION OF PROBLEM

2. Our first task is to decide on the functiontheoretic assumptions to be made in regard to F in (1). Naturally we would wish to avoid assumptions which might suggest that our investigation lacks finality.

Of course, we shall be working with a branch of w(x), analytic in some area in the plane of x. Pairing values of x and w gives a set of points $\mathfrak{M}$ in the space of w and x. It would be convenient to assume that F is of regular structure, as in VII, 3, at a point w = b, x = a of $\mathfrak{M}$ and that (1) is satisfied on $\mathfrak{M}$ for a neighborhood of (b, a). Here a critic might object. For instance, the equation

(2) $$(w - x)^{1/2} = 0$$

serves well to determine w as equal to x, and still the first member of (2) has a singularity wherever w = x. Should we not therefore admit equations (1) in which F has singularities on $\mathfrak{M}$? We are going to show that an equation (1) of greatest conceivable generality can be reduced to one in which F is of regular structure at a point (b, a) on $\mathfrak{M}$.

3. When, in what follows, we say that F(w, x) assumes the value zero at a point (w', x'), we shall mean that there exists an element of F with center at (w', x') which assumes there the value zero. There may be other elements of F with centers at (w', x') which are not zero at (w', x'). When we say that F has a singularity at (w', x'), we shall mean that there exists a sequence of elements of F, each an immediate continuation of its predecessor (as VII, 2), whose centers approach (w', x') and whose radii of convergence approach zero.

In specifying the nature of F, one will certainly assume that, (w', x') lying on $\mathfrak{M}$, F either assumes the value zero at (w', x') or has a singularity there. One will feel at ease if one is free to assume that F is zero at some points of $\mathfrak{M}$ and has singularities at the remaining points. One will feel the responsibility of explaining what it means for F to be zero at a singular point. Actually this last matter is unessential; the mere existence of the singularities will permit us to show that w(x) is elementary.

In our discussion, it will be convenient to employ an uncountable set of points of $\mathfrak{M}$ rather than all of $\mathfrak{M}$. We thus assume that there is an elementary function of w and x which is not identically zero and which satisfies at least one of the following two conditions:

(a) *It is zero at an uncountable set of points of* $\mathfrak{M}$.

(b) *It has singularities at an uncountable set of points of* $\mathfrak{M}$.

We shall show that there is an elementary function, not identically zero, of regular structure at a point (b, a) on $\mathfrak{M}$, which vanishes on $\mathfrak{M}$ in a neighborhood of (b, a).

4. From among all functions which satisfy (a) or (b), we select one of least order. We denote the function selected by F(w, x) and its order by n.

Suppose first that F is algebraic. Let F be defined by

(3) $$\alpha_0(w, x) u^p + \ldots + \alpha_p(w, x) = 0.$$

Wherever F is zero, $\alpha_p = 0$. Where F has a singularity, $\alpha_0 D$ with D the discriminant of (3) is zero. Thus $\alpha_0 \alpha_p D$ vanishes at an uncountable set of points of $\mathfrak{M}$. If we replace w in $\alpha_0 \alpha_p D$ by w(x), we get a function of x which clearly has to vanish identically in x. Thus $\alpha_0 \alpha_p D$ is a function such as we are seeking.

5. Let us suppose now that n > 0 and that F satisfies (b). We shall show that F can be replaced by a function of order n for which (a) holds.

F is given to us with regular structure at some point (b, a), not

necessarily on 𝔐. At (b, a), F has an element $P_0$. Let (w', x') be
one of the points on 𝔐 at which F has a singularity. Then there is
a curve C, joining (b, a) to (w', x') along which $P_0$ can be continued to any point preceding (w', x'), but not to (w', x') itself.
We form a chain of elements

$$P_0, P_1, \ldots, P_r, \ldots$$

with $P_{i+1}$ an immediate continuation of $P_i$, the center $(w_i, x_i)$ of
$P_i$ approaching (w', x') as i increases. Then the radius of convergence of $P_i$ approaches zero as i increases. Taking advantage of
fluency, we join (b, a) to a point $(\bar{w}_1, \bar{x}_1)$ close to $(w_1, x_1)$ by a
curve following C closely, the joining curve being such that each
complete monomial which appears in the structure of F can be continued along it. We take similarly a curve joining $(\bar{w}_1, \bar{x}_1)$ to a
point $(\bar{w}_2, \bar{x}_2)$ close to $(w_2, x_2)$. We continue in this manner, seeing to it that $(\bar{w}_i, \bar{x}_i)$ approaches (w', x') as i increases. We form
in this way a curve C', joining (b, a) to (w', x'), along which F
and each complete monomial can be continued to any point which precedes (w', x'). At $(\bar{w}_i, \bar{x}_i)$, the continuation of $P_0$ along C' will
be an element $Q_i$. We suppose each $(\bar{w}_i, \bar{x}_i)$ to be so close to
$(w_i, x_i)$ that $Q_i$ is an immediate continuation of $P_i$. Then the radius of convergence of $Q_i$ tends toward zero as i increases.

Let the equation which determines F at (b, a) be (𝔅) with each α
a polynomial in monomials. Two things may happen. Either one of the
complete monomials in the structure of F has a singularity at
(w', x') revealed by C', or else $\alpha_0$ D, where D is the discriminant
of (𝔅), assumes the value zero at (w', x'). We are going to show
that for an uncountable set of points (w', x') on 𝔐, $\alpha_0$ D is zero.
This will bring us back to the situation (a), because we have taken
(𝔅) so that D does not vanish identically.

6. Suppose it is not true that $\alpha_0$ D is zero at an uncountable set
of points of 𝔐. Then at least one of the complete monomials must
have singularities at an uncountable set of points of 𝔐. Let θ be
such a monomial. Let first θ = $e^v$ with v of order less than n. We
use (b, a) as above and a curve C joining it to a (w', x') on 𝔐 at
which θ has a singularity relative to C. We modify C slightly, as
in §5, so that v can be continued along it to any point which precedes (w', x'). We see that v has a singularity at (w', x'). This
contradicts the assumption that no function of order less than n
satisfies one of the conditions (a), (b). Suppose now that θ = log v.
Then v can be continued along C (modified if necessary), to any

point preceding $(w', x')$. If $v$ does not have a singularity at $(w', x')$, it must be that $v$ is zero there; otherwise $\theta$ could not have a singularity at $(w', x')$ relative to $C$. We have the contradiction that $v$ satisfies one of (a), (b). Thus $\alpha_0 D$ satisfies (a).

7. There is a function of order n which satisfies (a). We let F be such a function. We are going to show that F has a branch which is analytic at a point on 𝔐 and equal to zero on 𝔐 for a neighborhood of that point.

Let us consider the uncountable subset of points of 𝔐 at which F is zero. At each point F has an element, with center at the point, which is zero at its center. There must exist a positive $\varepsilon$ such that, for some uncountable subset E of these points, the radii of convergence of the elements of F exceed $\varepsilon$. If no such $\varepsilon$ existed, the elements at the centers of which F vanishes could be denumerated.

As is well known and easy to see, there exists some point $(d, c)$, every neighborhood of which contains an uncountable subset of points of E. Obviously we may assume that $(d, c)$ is on 𝔐 and that $w(x)$ is analytic for $|x - c| < \varepsilon$. Consider the neighborhood of $(d, c)$ given by

$$|w - d| < \frac{\varepsilon}{3}, \qquad |x - c| < \frac{\varepsilon}{3}$$

and the uncountable subset E' of E which lies in this neighborhood.

We now form for $(d, c)$ the immediate continuations of the elements of F with centers at the points of E'. The radius of convergence of each of these elements exceeds $2\varepsilon/3$, and each continuation equals zero at the center of the element from which it is obtained.

An infinite number of these immediate continuations must be identical. Otherwise there would be an uncountable set of distinct elements of F with centers at $(d, c)$ and we would have a contradiction of the theorem of Poincaré and Volterra which states that an analytic function can have at most a countable set of elements with the same center.*

It follows that one of the elements with center at $(d, c)$ vanishes for an infinite number of points of E'. Substituting for w

---

* This theorem is an immediate consequence of the fact that if $P_0$ is an element of F, and P a continuation of $P_0$, we can form a chain $P_0, P_1, \ldots, P_n = P$ with $P_{i+1}$ an immediate continuation of $P_i$, the elements $P_1, \ldots, P_{n+1}$ having centers at points $(w, x)$ at which, for w and x, the real part and the coefficient of i are rational.

in this element the function $w(x)$, we secure a function of x analytic at c which vanishes for an infinite number of points close to c and is therefore zero. This is the situation described at the beginning of the present section.

8. We shall now show that if F is suitably selected, there is a point on $\mathfrak{M}$, close to (d, c), at which F is of regular structure. We consider the equation (3) which defines F. We wish to show that we can replace F by $\alpha_p$ in (3). This will permit us to assume that F is a polynomial in monomials and will simplify the present discussion. Equation (3) holds at a point (b, a), not necessarily on $\mathfrak{M}$. Let (w', x') be any point on $\mathfrak{M}$ close to (d, c). There is a curve C, joining (b, a) to (w', x'), along which F is continued into the branch analytic at (d, c) secured above. We suppose C to be such that every complete monomial in the structure of F can be continued along it to any point preceding (w', x'). If there is a monomial which cannot be continued to (w', x'), that monomial has a singularity at (w', x'). As we saw above, this can happen only for a countable set of points (w', x'). We may therefore suppose, moving (d, c) if necessary, that each coefficient in (3) can be continued, together with F, from (b, a) to (d, c). Thus the continuation of $\alpha_p$ obtained at (d, c) vanishes on $\mathfrak{M}$ for a neighborhood of (d, c). We replace F by the m.a.f. of which $\alpha_p$ is a branch.

We now need only one detail to secure a point (d, c) on $\mathfrak{M}$ at which F is of regular structure. We use a curve C joining (b, a) to (d, c) along which every monomial can be continued. Every monomial in F has one of the forms $e^v$ or log v. As we saw in §6, we may suppose (d, c) and C to be such that the v for each monomial is continuable along C to (d, c). We consider the equations (3) for the various v. If $\alpha_0 D$ in each such equation is not zero at (d, c), F has regular structure at (d, c). Now, by the minimal nature of n, each $\alpha_0 D$ can vanish only for a countable set of points of $\mathfrak{M}$.

We may thus assume that $F(w, x)$ in (1) has regular structure at (d, c). In what follows, we write (b, a) for (d, c).

9. In discussing (1), we shall have to refer to those *monomials in the structure of F which involve w*. A complete monomial will be said to involve w if its partial derivative with respect to w is not identically zero. The structural scheme for F contains such a scheme for every monomial used in building F. We must arrange matters in such a way that if a monomial θ does not involve w, its structural scheme uses no monomials involving w. The reason is that we do not wish the differentiation of θ to introduce monomials involving w.

First, it is easy to see by induction that a monomial of any order n free of w is an elementary function of x in the sense of Chapter I; that is, it can be constructed without the use of the variable w. It will, in fact, be a monomial of order n as a function of x alone.

Now let the monomials free of w in the structure of F be $\theta_1, \ldots, \theta_p$. For each $\theta_i$ we secure a structural scheme as in I, 11, using a single point $a_1$ close to $\underline{a}$ as in (b, a). Shifting $\underline{a}$ if necessary, we assume that $a_1 = a$. We adjoin to the structure of F all monomials just secured for $\theta_1, \ldots, \theta_p$.

In the formal work below, most of what it is essential to bear in mind relative to F(w, x) is as follows. We have a finite number of monomials. F is algebraic in a certain subset of them. Some of them are free of w; those have derivatives which are algebraic in monomials free of w.

## COMPLETION OF PROOF

10. We suppose given an elementary F(w, x), not identically zero, of regular structure at (b, a) on $\mathfrak{M}$, which vanishes on $\mathfrak{M}$ in the neighborhood of (b, a). As we know, we may take F as a polynomial in monomials; we shall not insist on this.

We consider the class A of all such functions; (b, a) may be different for different functions. We suppose that each function has a structure in which the maximum, r, of the orders of those monomials which involve w is as low as possible. We assume further that s, the number of r-monomials involving w, is as small as possible.

We consider those functions in A for which r has a least value, say $r_0$. From this subset we choose one, F, for which s has a minimum value $s_0$. Let $\theta$ be one of the $r_0$-monomials in F which involve w. We write

(4) $$F(w, x) = G(\theta, w, x).$$

The successive partial derivatives of G with respect to $\theta$ cannot all vanish at (b, a). If they did, we could replace $\theta$ in (4) by $\theta(b, a)$ and $s_0$ would not be a minimum. Suppose then that the first j derivatives in $\theta$ vanish on $\mathfrak{M}$ for a neighborhood of (b, a), but that the derivative of order j + 1 is distinct from zero at some point close to (b, a). We assume, as we may, that this point is (b, a) itself. We designate the $j^{th}$ derivative by H and write, for the w(x) of §1,

(5) $$H[\theta(w, x), w, x] = 0.$$

For the neighborhood of $(b, a)$, we may solve for $\theta$ in (5) and write

(6) $$\theta = f(w, x).$$

If $r_0 = 0$, $\theta$ is $w$ and (6) states that $w(x)$ is an elementary function of $x$. In what follows, we suppose that $r_0 > 0$.

Suppose that $\theta = e^v$ with $v$ of order $r_0 - 1$. For $w$ and $x$ independent variables close to $(b, a)$ we write (6) in the form

(7) $$v(w, x) = \log f(w, x).$$

Now, letting $w$ in (7) be $w(x)$, we find, differentiating with respect of $x$,

(8) $$v_w(w, x) y(x) + v_x(w, x) = \frac{f_w(w, x) y(x) + f_x(w, x)}{f(w, x)}.$$

In (8) $y(x)$ is the function of which $w(x)$ is an integral. We suppose, as we may, that $y$ has regular structure at $\underline{a}$.

Now (8) must be an identity in $w$ and $x$ for a neighborhood of $(b, a)$. Otherwise it would give a function in $A$ with $r \leq r_0$ and with $s < s_0$ if $r = r_0$.

The two functions

(9) $$v(w + \mu, x), \qquad \log f(w + \mu, x)$$

are analytic in $w$, $x$, $\mu$ for $w = b$, $x = a$, $\mu = 0$.

Let us imagine that in (9) we replace $w$ by $w(x)$. The two functions of (9) become functions of $x$ whose derivatives are given by the two members of (8) with $w$ replaced by $w(x) + \mu$. We therefore have

(10) $$v[w(x) + \mu, x] = \log f[w(x) + \mu, x] + \beta(\mu)$$

with $\beta$ analytic at $\mu = 0$. Then

(11) $$v_w(w, x) = \frac{f_w(w, x)}{f(w, x)} + c$$

for a neighborhood of $(b, a)$ on $\mathfrak{M}$, and therefore identically in $w$ and $x$. Integrating, we have in $w$ and $x$

(12) $$v(w, x) = \log f(w, x) + cw + \gamma(x).$$

To determine $\gamma(x)$ we consider a $(b', a')$ close to $(b, a)$ for which (7) does not hold. For $w = b'$, (12) gives

(13) $\qquad \gamma(x) = v(b', x) - \log f(b', x) - c b'.$

By (7) and (12) we have for $w = w(x)$,

(14) $\qquad\qquad\qquad c w = - \gamma(x).$

Now c is not zero. If it were, (13) would imply that

$$v(b', x) - \log f(b', x)$$

vanishes for every x, whereas it is not zero for $x = a'$.

By (14), $w(x)$ is elementary.

If $\theta = \log v$, (6) becomes

$$f(w, x) = \log v(w, x)$$

and we find again that $w(x)$ is elementary.

We might have used, in what precedes, a different classification of functions of w and x, letting w, and every elementary function of x alone, be monomials of order zero. This would have spared us the considerations relative to monomials involving w. Having already classified our functions in Chapter VII, we found it easier to proceed as above. A classification of the new type will be used below in connection with Bessel's equation.

In the introduction to my paper, [18], I mentioned the problem of determining whether the integral of an elementary function is elementary if it is one of a set of functions which satisfy a set of elementary equations. I stated that the formal elements of the proof for one function could be carried over to answer this question affirmatively. I wish to withdraw this statement. I do not have the details now, if indeed I ever did. In particular, the question as to whether the integral of an elementary function may be represented parametrically by elementary expressions may be of interest.

## *LINEAR DIFFERENTIAL EQUATIONS OF THE SECOND ORDER*

11. In Chapter VI, we determined when a solution of Bessel's equation is a function of Liouville. We shall now examine the question of implicit representations.

We shall construct Liouville functions of w and x, letting w and every Liouville function of x be complete monomials of order 0. A function of order 0 is an algebraic combination of complete monomials of order 0. An m.a.f. $u(w, x)$ is a complete monomial of order 1 if it is not a function of order 0 and if either

(a) $u = e^v$ with $v$ of order $0$

or

(b) $u_w$ and $u_x$ are both functions of order $0$.

The construction continues in the usual way.

12. The differential equation

$$y'' + 2\,P(x)\,y' + Q(x)\,y = 0$$

goes over, under the transformation $y = e^{-\int P\,dx}$, into an equation

(15) $$w'' = \varphi(x)\,w.$$

Accordingly, we deal with an equation (15). We prove the theorem.*

THEOREM: *If $w$, a solution of the equation $w'' = \varphi(x)\,w$, in which $\varphi(x)$ is a Liouville function, satisfies an equation $F(w, x) = 0$, where $F$ is a Liouville function not identically zero, then $w$ is a Liouville function of $x$.*

We proceed formally; one has seen how rigor is secured.

Differentiating $F = 0$ with respect to $x$, we find

(16) $$F_w\,w' + F_x = 0.$$

The first member of (16) is a function algebraic in $w'$ and in Liouville monomials in $w$ and $x$ which vanishes when $w$ is the given solution of (15). Of course $F$ is also such a function, but (16) gives a better example.

With every expression for every such function we associate two numbers, first $r$, the maximum of the orders of the monomials in $w$ and $x$, and secondly $s$, the number of $r$-monomials. We take those expressions for which $r$ is a minimum and select from them an $f(w', w, x)$ for which $s$ is a minimum. We write, for the given solution of (15),

(17) $$f(w', w, x) = 0$$

and denote the two integers for the first member of (17) by $r$ and $s$.

13. We are going to prove that $r = 0$. Let $r > 0$ and let $\theta$ be one of the $r$-monomials in $f$. Solving (17) for $\theta$, we find

(18) $$\theta - g(w', w, x) = 0$$

where $g$ is algebraic in $w'$ and monomials and has $s - 1$ monomials of order $r$.

If $\theta = e^v$ with $v$ of order $r - 1$, we write

* Ritt [21]. I cannot agree with a remark of Mordukhai-Boltovskoi [13] to the effect that the question now being considered is equivalent to one previously settled by him. The two questions seem quite distinct and not reducible to each other in any obvious way.

(19) $$v - \log g(w', w, x) = 0.$$

We let the first member of

(20) $$G(w', w, x) = 0$$

represent the first member of (18), or that of (19), according as θ is an integral or an exponential.

Differentiating (20) and using (15), we have

(21) $$\varphi(x) \, w \, G_{w'} + G_w \, w' + G_x = 0$$

which holds identically in $w'$, $w$, $x$.

Let $\mu$ be any constant. The derivative with respect to $x$ of

$$G(\mu w', \mu w, x)$$

will be the first member of (21) with $w'$ and $w$ replaced by $\mu w'$ and $\mu w$. Thus

$$G(\mu w', \mu w, x) = \beta(\mu).$$

We differentiate with respect to $\mu$ and put $\mu = 1$. Then

(22) $$w' \, G_{w'} + w \, G_w = c$$

identically in $w'$, $w$, $x$. A particular solution of (22) is $c \log w$, and $w'/w$ is a solution of the equation obtained from (22) by putting $c = 0$. Then

(23) $$G = c \log w + H\left(\frac{w'}{w}, x\right).$$

For the given solution of (15), we have

(24) $$c \log w + H\left(\frac{w'}{w}, x\right) = 0.$$

We put $u = w'/w$ and differentiate with respect to $x$. Then

(25) $$c \, u + [\varphi(x) - u^2] \, H_u + H_x = 0.$$

14. Let θ be an integral. As, in this case, G is algebraic in $w'$, H must be algebraic in $u$ and in Liouville monomials in $x$. Then (25) is an identity in $u$ and $x$. From (23), we see that $c \neq 0$. We shall show that no H which is algebraic in $u$ can satisfy (25), thus making untenable, in the case in which θ is an integral, the assumption that $r > 0$. Let one of the developments of H about $u = \infty$ be

(26) $$\alpha_1(x) \, u^{p_1} + \alpha_2(x) \, u^{p_2} + \ldots$$

with no $\alpha$ zero. Let $p_1 \ne 0$. The expansion of $(\varphi - u^2) H_u$ begins with an exponent $p_1 + 1$, whereas in $H_x$ the first exponent does not exceed $p_1$. Hence the first term of $(\varphi - u^2) H_u$ must balance with $c u$, an impossibility when $p_1 \ne 0$. If $p_1 = 0$, $(\varphi - u^2) H_u$ has $p_2 + 1$ for first exponent, and $H_x$ begins with a zero or negative power. As $c \ne 0$, we have to balance $c u$ in (25). This is impossible because $p_2 + 1 < 1$. Thus no algebraic H satisfies (25).

15. Let $\theta$ be an exponential. Then $e^G$ is algebraic in $w'$. Then, if K represents $e^H$ with H as in (23), K is algebraic in $w'/w$. We find from (25),

(27) $$c u K + (\varphi - u^2) K_u + K_x = 0.$$

We consider an expansion (26) for K, and substitute it into (27). The first equation obtained is $c - p_1 = 0$. By (24) we have, for the given solution of (15),

(28) $$w^{p_1} K(\frac{w'}{w}, x) = 1.$$

As (28) is algebraic in $w'$ and $w$, we have a final contradiction of the assumption that $r > 0$.

16. We consider (17) with f algebraic in $w'$ and w. If (17) does not involve $w'$, it determines $w(x)$ as a Liouville function of x. If (17) does involve $w'$, we find

(29) $$w' = H(w, x)$$

with H algebraic in w and Liouville monomials in x.

Differentiating (29), we have

(30) $$\varphi(x) w = H_w H + H_x.$$

If (30) is not an identity in w and x, it determines w as a Liouville function. Let (30) be an identity. We consider a development of H about $w = \infty$,

$$H = \alpha_1(x) w^{p_1} + \ldots.$$

The $\alpha$ are functions of Liouville. When we substitute into (30) we find that $p_1 = 1$ and that

$$\alpha_1' + \alpha_1^2 = \varphi.$$

If we let u be the exponential of the integral of $\alpha_1$, we find that $u'' = \varphi(x) u$. Thus u, a nonzero Liouville function, satisfies (15).

The knowledge of one nonzero solution of (15) permits us to reduce (15) to a linear equation of the first order and to get its complete solution, with two constants, by quadratures. This completes the proof of our theorem.

## REFERENCES*

[1] Abel, N. H. *Oeuvres complètes*. See also Serret, *Cours d'algèbre supérieure*, Vol. II. Paris, 1928.

[2] Koenigsberger, L. "Bemerkungen zu Liouville's Classificirung der Transcendenten," *Mathematische Annalen*, XXVIII (1887), 482-92.

[3] Liouville, J. "Sur la détermination des intégrales dont la valeur est algébrique," *Journal de l'école polytechnique*, XIV (1833), Section 23, 124-93.

[4] ————"Mémoire sur les transcendantes elliptiques de première et de seconde espèce, considerées comme fonctions de leur amplitude," *Journal de l'école polytechnique*, XIV (1833), Section 24, 57-83.

[5] ————"Mémoire sur l'intégration d'une classe de fonctions transcendantes," *Journal für die reine und angewandte Mathematik*, XIII (1835), 93-118.

[6] ————"Mémoire sur la classification des transcendantes et sur l'impossibilité d'exprimer les racines de certaines équations en fonction finie explicite des coefficients," *Journal de mathématiques, pures et appliquées*, II (1837), 56-104; III, 523-46.

[7] ————"Mémoire sur l'intégration d'une classe d'équations différentielles du second ordre en quantités finies explicites," *Journal de mathématiques, pures et appliquées*, IV (1839), 423-56.

[8] ————"Mémoire sur les transcendantes elliptiques de première et de seconde espèce, considerées comme fonction de leur module," *Journal de mathématiques, pures et appliquées*, V (1840), 34-36, 441-64.

[9] ————"Remarques nouvelles sur l'équation de Riccati," *Journal de mathématiques, pures et appliquées*, VI (1841), 1-13.

[10] Lorch, E. R. "Elementary Transformations," *Annals of Mathematics*, XXXIII (1932), 214-28.

[11] Maximovich, W. P. "Determination of the General Equation of the First Order Which Can Be Integrated in Finite Terms."

* This is not a complete list of publications on integration in finite terms.

## REFERENCES

Kasan, 1885. (In Russian). See also *Comptes Rendus de l'Académie des Sciences de Paris*, CI (1885), 809-11.

[12] Mordukhai-Boltovskoi, D. *On the Integration in Finite Terms of Linear Differential Equations*. Warsaw, 1910. (In Russian)

[13] ———— *On the Integration of Transcendental Functions*. Warsaw, 1913. (In Russian)

[14] ———— "Researches on the Integration in Finite Terms of Differential Equations of the First Order," *Communications de la société mathématique de Kharkov*, X (1906-1909), 34-64, 231-69. (In Russian)

[15] ———— "Sur la résolution des équations différentielles du premier ordre en forme finie," *Rendiconti del circolo matematico di Palermo*, LXI (1937), 49-72.

[16] Ostrowski, A. "Sur les relations algébriques entre les intégrales indéfinies," *Acta Mathematica*, LXXVIII (1946), 315-18.

[17] ———— "Sur l'intégrabilité élémentaire de quelques classes d'expressions," *Commentari Mathematici Helvetici*, XVIII (1946), 283-308.

[18] Ritt, J. F. "On the Integrals of Elementary Functions," *Transactions of the American Mathematical Society*, XXV, (1923), 211-22.

[19] ———— "Elementary Functions and Their Inverses," *Transactions of the American Mathematical Society*, XXVII (1925), 68-90.

[20] ———— "Simplification de la méthode de Liouville dans la théorie des fonctions élémentaires," *Comptes Rendus de l'Académie des Sciences de Paris*, CLXXXIII (1926), 331-32.

[21] ———— "On the Integration in Finite Terms of Linear Differential Equations of the Second Order," *Bulletin of the American Mathematical Society*, XXXIII (1927), 51-57.

[22] ———— "Algebraic Combinations of Exponentials," *Transactions of the American Mathematical Society*, XXXI (1929), 654-79.

[23] Watson, G. N. *Bessel Functions*. Cambridge, 1944, pp. 111-23.

Bei Fragen zur Produktsicherheit wenden Sie sich bitte an:
If you have any questions regarding product safety,
please contact:

Walter de Gruyter GmbH
Genthiner Straße 13
10785 Berlin
productsafety@degruyterbrill.com